Mechanical Engineering Series

Series Editor

Vish Prasad, Department of Mechanical and Energy Engineering
University of North Texas
Denton, USA

The Mechanical Engineering Series presents advanced level treatment of topics on the cutting edge of mechanical engineering. Designed for use by students, researchers and practicing engineers, the series presents modern developments in mechanical engineering and its innovative applications in applied mechanics, bioengineering, dynamic systems and control, energy, energy conversion and energy systems, fluid mechanics and fluid machinery, heat and mass transfer, manufacturing science and technology, mechanical design, mechanics of materials, micro- and nano-science technology, thermal physics, tribology, and vibration and acoustics. The series features graduate-level texts, professional books, and research monographs in key engineering science concentrations.

Zainul Huda

Metal Casting Engineering

Design, Processes, Calculations

 Springer

Zainul Huda
Professor emeritus, Mechanical Engineering
King Abdulaziz University
Jeddah, Saudi Arabia

ISSN 0941-5122   ISSN 2192-063X   (electronic)
Mechanical Engineering Series
ISBN 978-3-031-84622-9   ISBN 978-3-031-84620-5   (eBook)
https://doi.org/10.1007/978-3-031-84620-5

This Springer imprint is published by the registered company Springer Nature Switzerland AG
The registered company address is: Gewerbestrasse 11, 6330 Cham, Switzerland

If disposing of this product, please recycle the paper.

Preface

This is a unique textbook featuring the principles, design, practices, and analyses of metal casting processes supported by 200 worked numerical examples. *Metal Casting* or *Foundry Technology* course is offered in BS (Metallurgical/Mechanical Engineering) programs at reputed universities. Metal Casting is also taught as an essential part of Manufacturing Technology/Processes core course in BS (Manufacturing/Mechanical/Industrial Engineering) programs in almost all universities. Besides UG students, graduate/PG students and researchers as well as practicing engineers can benefit from this book because this volume covers both traditional casting design and processes and the recent advances in metal casting technology.

This volume covers all important aspects of metal-casting processes and practices. Foundry practices are explained with the aid of 120 labeled illustrations. One of the salient features of the book is the inclusion of an industrially-oriented casting-design project. The key solution to the project is presented with the aid of mathematical analysis and diagrams. There are 100 multiple-choice questions (MCQs), whose answers are given at the end of the book along with the answers to selected problems.

There are thirteen (13) chapters in this book, which is composed of an introductory chapter and three (3) parts. Part I (Chaps. 2, 3, 4, and 5) deals with melting and solidification design. Chapter 2 explains the melting and flow of metals, with particular reference to fluidity, *Bernoulli's law,* and pouring kinetics. Chapter 3 covers solidification design with reference to Fick's law of heat transfer and *Chvorinov's rule;* here the cooling curve and the effects of solidification kinetics on the grain size, casting's strength, and micro-segregation are also discussed. Chapter 4 describes melting practices and furnaces, including the melting stock, mixing ratio, and their effects on the environment. This chapter also presents the melting furnace construction, design, working principle, charge calculations in electric arc furnace (EAF), induction furnace, and crucible furnace. Chapter 5 explains the cold-blast cupola covering its working principle, construction, design analysis, and charge calculations; here engineering analysis for the optimum blast flow rate to achieve the maximum metal superheating is also presented.

Part II (Chaps. 6, 7, 8, 9, 10, and 11) presents casting design and processes. Chapter 6 discusses the permanent-pattern expendable-mold casting (PPEMC) processes, including the properties of foundry sand and pattern making; here, the discussion on PPEMC processes includes green sand mold casting process and other sand-water-clay bond casting processes, resin-bond casting processes (shell mold casting, hot-box casting, cold-box casting), silicate-bond casting, and V-casting processes. Chapter 7 presents sand casting design, including the designs for pattern, gating system (sprue, runner, etc.), riser, and mold; here recent developments in the designs of sprue and riser are also included. Chapter 8 develops, in readers, the project-engineering skill through solving an industrially-oriented casting-design project. This chapter presents a worked casting-design project; here, the engineering drawing of a component (to be cast) is given, and it is required to design sprue, riser, and the mold to produce a defect-free casting. Chapter 9 explains expendable-pattern expendable-mold casting (EPEMC) processes with reference to the investment casting process (ICP), lost foam casting (LFC), and full mold casting, including their engineering analyses and recent developments. Chapter 10 focuses on the technological and economic aspects of die casting processes. These processes include gravity die casting (GDC), low-pressure die casting (LPDC), and high-pressure die casting (HPDC) (both hot-chamber and cold-chamber HPDC machines). The engineering analyses of GDC and HPDC are presented with reference to the determinations of casting cost and the cost of a unit HPDC product. Chapter 11 describes centrifugal casting processes covering true centrifugal casting (TCC), semi-centrifugal casting (SCC), and centrifuge casting; here, engineering analysis of TCC includes the derivation of a mathematical expression for the mold rotational speed, and the design of a transmission system from motor to centrifugal casting machine.

Part III (Chaps. 12 and 13) considers the quality, safety, and lean manufacturing aspects in foundry practice. Chapter 12 presents an exhaustive treatment of mechanical testing, casting defects, and quality assurance (QA) in metal casting. Mechanical testing covers both destructive testing (tensile-, hardness-, impact-, and fatigue testing) and nondestructive testing (NDT, e.g., magnetic particle testing, dye-penetrant testing, ultrasonic testing, and radiography). Sand-casting defects (misrun, hot tearing, shrinkage cavity, blow holes, pin-holes, metal penetration, mismatch) are explained, and their remedies are presented for QA. The avoidance of porosity defect in aluminum die castings is also discussed with reference to the latest vacuum technology. Chapter 13 emphasizes the safety and lean manufacturing aspects in a foundry; here, foundry safety is discussed with reference to hazard assessment, personal protective equipment (PPE), environmental regulations, heat stress, the moisture issue, machinery inspection, dust exposure, noise pollution, and harmful vibrations. Lean manufacturing is introduced both in general and particularly with reference to lean practices in foundries; here, a case study of waste-minimization in a sand-casting foundry is also included. Lean manufacturing in a die-casting (HPDC) foundry is also considered, covering material selection, the design for manufacturability (*DfM*), production volumes and economy, tooling and die cost, and reducing rejections.

Based on my 42 years' experience in the fields of metallurgical/manufacturing engineering in academia/industry, I hope this volume will be highly beneficial for engineering academicians, students, and practicing engineers in foundries.

Jeddah, Saudi Arabia   Zainul Huda
December, 2024

Acknowledgments

All thanks to God who blessed me with wisdom to complete the write-up of this comprehensive book. I am grateful to my wife for cooperating with me in writing the book at nights. I would like to acknowledge Greg Bray, President, EC&S Inc., Birmingham AL, USA, for his permission to publish the picture of the cupola in his foundry. I am grateful to Tuan Zaharinie, PhD (Manufacturing Engineering), Associate Professor, Mechanical Engineering Department, University of Malaya, Kuala Lumpur, Malaysia, for reviewing Chap. 7 and for her valuable suggestions to improve the contents of the chapter. I also like to thank Jamal Husain Qadri, PhD (Materials Engineering), President, Solid Solutions International LCC, Sahuarita AZ 85629, Arizona, USA, for reviewing Chap. 3 and providing healthy suggestions for improvement. Finally, I acknowledge Adil A. Ashary, PhD, Distinguished Engineer, Bloom Energy Corporation, San Jose, California, USA, for partially reviewing Chap. 2.

Contents

Part III Safety, Quality, and Lean Manufacturing

About the Author

Zainul Huda is a Professor emeritus of Manufacturing/Mechanical Engineering, King Abdulaziz University, Jeddah, Saudi Arabia. He has worked as a full Professor in good standing at reputed/world-ranking universities for 17 years (February 2007–May 2024). Professor emeritus Zainul Huda possesses 42 years' academic experience in manufacturing/mechanical/metallurgical engineering. His teaching interests include Foundry/Manufacturing Technology, Metallurgy, Fracture Mechanics, Materials Science/Engineering, and Mechanical Behavior of Materials. He has supervised to completion several PhD and Masters' theses in the field of manufacturing and materials engineering.

Professor emeritus Z. Huda is the single/first author of 9 books and over 100 research articles in the field of metallurgical/materials/manufacturing engineering published by reputed journals/publishers from the USA, Canada, UK, Germany, France, Poland, Switzerland, Pakistan, South Korea, Malaysia, and Singapore. His research work has been cited over 1380 times with h-index of 17 and i-10 index of 29. He (as a PI) has attracted nine (9) research grants, with a total worth of US\$ 0.16 million.

Professor emeritus Zainul Huda has hands-on experience in metal casting. He has worked as a manager/engineer/metallurgist in various manufacturing companies, including Pakistan Steel Mills Corporation Ltd., Karachi. He is a professional engineer (PE) registered under the Pakistan Engineering Council. He has also successfully completed 20 industrial consultancy projects in the areas of failure analysis and manufacturing. He is the developer of Toyota Corolla cars' axle-hub's manufacturing procedure first ever implemented in Pakistan (through Indus Motor Co., Karachi).

Professor emeritus Z. Huda earned a PhD in Materials Engineering/Metallurgy from Brunel University, London, UK, in 1991. He is also a postgraduate in Manufacturing Engineering. He obtained a BS (Metallurgical Engineering) from the University of Karachi, Pakistan, in 1976. He has delivered guest lectures on metallurgy/mechanical engineering in the United Kingdom, South Africa, Malaysia, Pakistan, and Saudi Arabia. Prof. em. Huda is the recipient of UK-Singapore Partners in Science Collaboration Award, which was awarded to him by the British High Commission, Singapore in 2006. He is also the recipient of the University of Malaya's Vice Chancellor's Appreciation Certificate for an Innovation in the engineering design and fabrication of a machine.

Chapter 1
Introduction

Abbreviations

ADI	Austempered ductile iron
CC	Continuous casting
CG	Columnar grains
DS	Directional solidification
EPEMC	Expendable-pattern expendable-mold casting
LMC	Liquid metal cooling
PMC	Permanent mold casting
PPEMC	Permanent-pattern expendable-mold casting
QC	Quality control
SC	Single crystal

1.1 Metal Casting and Its Industrial Importance

1.1.1 What Is Metal Casting?

Metal casting is the process of manufacturing objects by pouring molten metal into an empty shaped space—*mold cavity*. This process basically involves mold preparation, melting of metal, and pouring it to fill the mold cavity; the poured metal retains the desired shape of the mold cavity after solidification. Casting is a cost-effective manufacturing process, since it is a less expensive way to produce a part as compared to machining. There are a wide variety of different metal casting processes (see Sect. 1.3); however, sand casting is the most economical method to manufacture a cast product (see Fig. 1.1). Sand casting involves the use of a sand mold to manufacture a part by metal casting. A detailed account of sand casting is given in Chaps. 6 and 7.

© The Author(s), under exclusive license to Springer Nature
Switzerland AG 2025
Z. Huda, *Metal Casting Engineering*, Mechanical Engineering Series,
https://doi.org/10.1007/978-3-031-84620-5_1

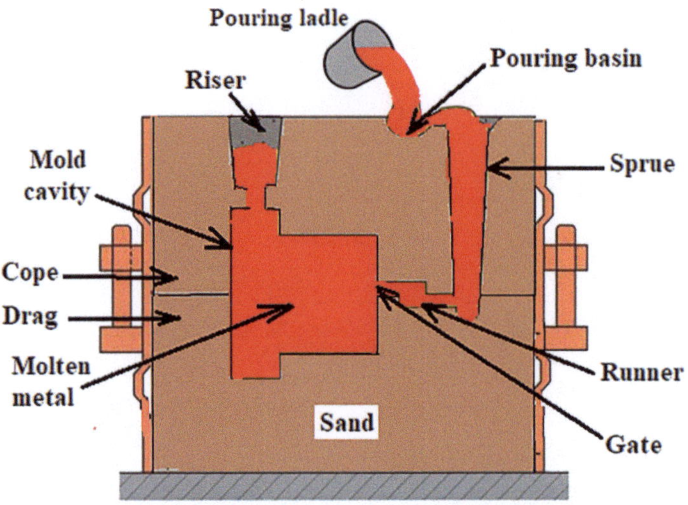

Fig. 1.1 A schematic of sand casting process—pouring of molten metal into a sand mold

1.1.2 Metal Casting and Foundry: Advantages and Importance

There are several distinct advantages of metal casting over other manufacturing processes. These advantages include: (a) the creation of complex part geometries, (b) production of both external and internal shapes, (c) economy of casting process, with little wastage (the extra metal is re-melted and re-used), (d) some casting processes are *net shape* or near net shape, and (e) the production of a wide variety of sized parts (e.g., large parts: engine blocks, cylinder heads, railway wheels, pipes; and small parts: dental crowns, jewelry, gears, brake components, etc.).

The metal casting process is carried out in a foundry, which has facilities for safety, metals/alloys storage, furnace(s), mold preparation materials and equipment, molten-metal pouring, removal of casting, cleaning, inspection, and quality assurance. Metal castings (cast products) find broad applications in industrial machinery, automotive components, aerospace components, defense components, and many other engineering applications. Today, foundries are a $33 billion dollar industry in the United States. There are over 2500 foundries in the United States today and tens of thousands around the world (Quaker City Castings, 2024).

1.1.3 Industrial Applications of Metallic Cast Products

Our modern lives are filled with cast metal products. Some notable examples of commonly used metallic cast products include: engine housings, machinery parts, automotive parts, gas-turbine blades, train wheels, trailer hitches, lamp posts, and

Table 1.1 Some typical cast products, their materials, and applications

Products/machine elements	Usual materials	Casting process	Applications
Gas-turbine blades	Superalloy (single crystal or columnar grained)	Directional solidification	Gas-turbines/ aero- engines
Cylinder liners	Cast iron, other alloys	Chill casting	Compressors and engines
Engine housings	Aluminum alloy	Die casting	I.C. Engines
Gear-box housing	Aluminum alloy	Die casting	Aircraft
Gears, crankshaft	Nodular cast iron, Austempered ductile iron (ADI)	Shell molding, investment casting, etc.	Automotive, machinery
Thin-walled cylinders	Suitably selected metal/ alloy	Centrifugal casting	Petro-chemical plants, chemical plants, etc.
Door stopper	Zinc	Die casting	Door stopping
Slabs, blooms	Steels	Continuous casting	For rolled products
General-purpose machine elements	Suitably selected metal/ alloy	Sand casting	General engineering applications

the like. For example, aluminum-alloy engine housings and gearbox housings, for automotive and aerospace applications, are manufactured by die casting. Thin-walled cylinders, for applications in chemical/petro-chemical plants, are produced by centrifugal casting. Superalloy single-crystal turbine-blades are manufactured by directional-solidification (DS) casting process (Huda, 2020). In fact, many high-quality machine components are manufactured by metal casting processes. Some typical metallic cast products, their materials, usual casting process, and applications are presented in Table 1.1. Additionally, some sand cast and die cast machine components are shown in Fig. 1.2.

1.2 Basic Requirements and Principal Steps in Metal Casting

1.2.1 Basic Requirements in Metal Casting

It is learnt in the preceding section that metal casting process is carried out in a foundry, which has facilities for safety, furnace(s) for melting the metal/alloy, mold preparation materials and equipment, molten-metal pouring, removal of casting, cleaning, and inspection. It is, therefore, important for a foundry-person to ensure the availability of basic requirements/facilities prior to starting a casting process. There are seven basic requirements/facilities for carrying out the metal casting process: (1) mold preparation facility, (2) melting furnace, (3) pouring facility, (4) solidification design, (5) mold removal facility, (6) cleaning, finishing, inspection, and quality assurance (QC), and (7) safety and environmental protection facility. These casting requirements are explained in the following paragraph.

Fig. 1.2 Some metallic cast products/machine elements

A *mold* must be produced with a mold cavity with the desired shape and size of the solidified metal. A *melting furnace* must be available with the capability of providing molten metal at the proper temperature in the desired quantity and quality; there must be a proper chimney system to avoid pollution and to ensure a safe and *clean environment* in the foundry. A *pouring ladle/facility* must be available to receive the molten metal from the furnace, and to introduce the molten metal into the mold to fill the mold cavity. The *solidification process* should be properly analyzed and designed to produce defect-free castings. A mold removal technique must be devised to properly remove the casting from the mold. Tools and equipment must also be available for *cleaning, finishing, and inspecting* the cast product (see Chap. 12). The *safety* of workers and the protection of a clean environment are of utmost importance in foundry practice; these considerations are discussed in Chap. 13.

1.2.2   Principal Steps in Metal Casting

The following principal steps must be followed for carrying out the sand casting process: (a) making of the master pattern and core box (if required), (b) preparation of the mold and core (if needed), (c) melting of the metal/alloy, (d) pouring of the metal into the mold, (e) allowing the metal to solidify, (f) removing the casting from the mold, and (g) finishing the casting and its heat treatment (if required) as per quality standards. Each of these steps must be properly designed and performed to avoid casting defects (such as porosity, shrinkage cavity, blow holes, etc.). The principal steps in metal casting are illustrated in the flow chart (Fig. 1.3), and described in the following paragraphs.

Making the Master Pattern and Core Box   The first step in sand casting is to make the *master pattern*—an object with a similar shape to that of the desired product.

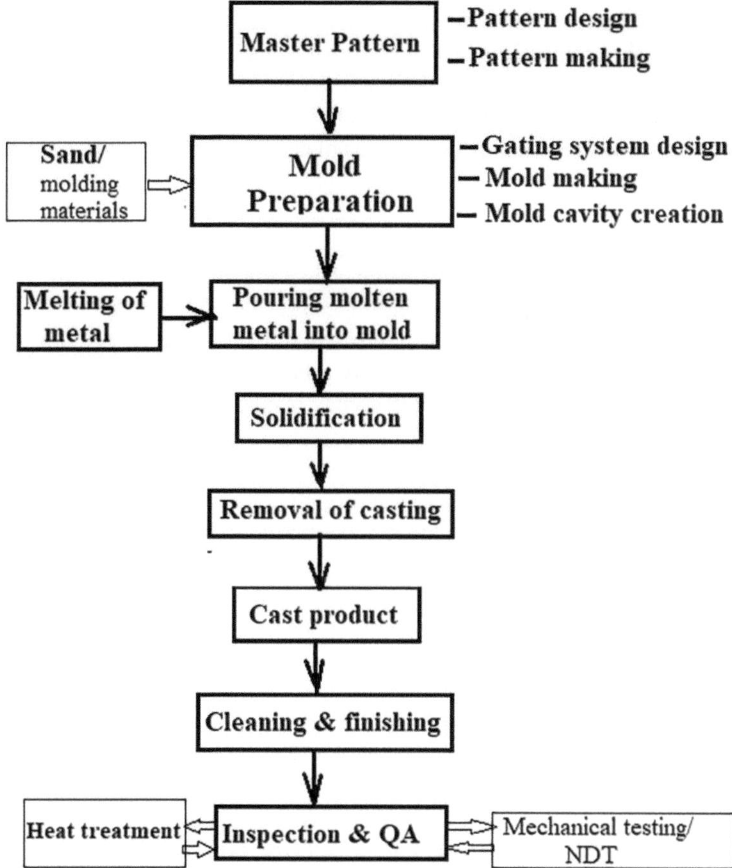

Fig. 1.3 The principal steps in sand casting

The pattern is a tool used to produce the mold, and must be over-sized than the final product in order to account for shrinkage during solidification. The pattern can be produced either by machining or by 3D printing/additive manufacturing. Similarly, *cores* are made using a *core-box*—another type of tooling. Permanent-mold casting processes (utilizing reusable molds) do not normally require patterns or core-boxes. In such cases, toolmakers can directly produce the mold and core.

Preparing the Mold and the Core Successful manufacturing of a casting (cast product) requires preparation of a mold with a mold cavity; the latter contains and forms the molten metal into the desired shape (see Fig. 1.1). In many cases, there are holes or empty spaces in the casting geometric design. In such cases, the manufacturer must also produce a *core* that fits inside the mold to form hollow cavities inside the final part. Molds and cores are either reusable (as in the case of permanent mold casting), or single-use (as in expendable mold casting).

Melting the Metal/Alloy It is necessary to melt the metal or alloy in a furnace, which must be capable of heating the metal and melt it to a temperature reasonably above its melting temperature. Foundry furnaces include cold-blast cupola, induction furnace, electric arc furnace, crucible furnace, and the like (see Chaps. 4 and 5). It is important to heat the metal to a temperature (pouring temperature) that is reasonably higher than the melting point of the metal. The difference between the melting temperature and the pouring temperature is called *superheat* (see Chap. 3, Sect. 3.3).

Pouring the Molten Metal into the Mold Next, the molten metal must be poured into the mold by the use of a *pouring ladle* (see Fig. 1.1). Modern ladles are designed with facilities that support the pouring process and allow operators to control pouring speed either manually or automatically. Improper pouring techniques can lead to defects; for example, a pin-hole defect may result in a casting due to trapping of gases inside the mold.

Solidification On the completion of pouring, the metal must be allowed to solidify within the mold for the pre-determined time duration; the metal in the mold cavity takes the shape of the casting. The solidification time can be calculated using *Chvorinov's rule* (see Chap. 3). It is also technologically important to consider shrinkage of the molten metal as it cools, as well as release of gases that build up inside the mold during pouring.

Removal of the Casting from the Mold Once the metal has solidified and cooled, the casting must be removed from the mold. In the case of a method using expendable molds, the worker may simply break open the mold to remove the casting. For permanent molds, lubricating coatings are often utilized to prevent sticking, and features allowing casting removal (e.g., ejector pins) must be provided into the mold.

Inspection and QC Visual inspection usually requires additional finishing processes, such as cleaning, gate removal, and riser removal. In order to meet quality control (QC) specifications, mechanical testing and metallographic inspection must be conducted; the latter may require machining and heat treatment.

1.3　Metal Casting Processes: Classification and Overview

A variety of casting processes/methods are used in foundries. Figure 1.4a–d presents the comprehensive classification chart of various metal casting processes. All casting processes can be broadly classified into two groups: (1) expendable mold casting (EMC) processes and (2) permanent mold casting (PMC) processes. In addition to the two groups of casting processes, there are other non-traditional methods, such as chill casting, continuous casting, etc. (see Fig. 1.4a). Expendable-mold casting and permanent-mold casting processes are briefly described in the following section (Sect. 1.4). These casting processes are discussed in detail in Chaps. 6, 7, 8, 9, 10 and 11. Chill casting and continuous casting processes are explained in Sect. 1.5.

1.4　Expendable-Mold and Permanent-Mold Casting Processes

Expendable-mold casting processes are based on single-use molds. These methods involve the use of temporary, non-reusable molds (see Part III, Chaps. 6, 7, 8 and 9). These processes can be sub-divided into two groups: (a) permanent-pattern expendable-mold casting (PPEMC) processes and (b) expendable-pattern expendable-mold casting (EPEMC) processes. Examples of PPEMC processes are based on the following types of bond in the molding materials: (a) sand-, water-, and clay bond, (b) resin bond, (c) silicate bond, and (d) no bond (see Fig. 1.4b). Examples of sand-water-and-clay bond PPEMC processes include green-sand mold casting, dry-sand mold casting, floor mold casting, pit mold casting, and loam mold casting processes (see Fig. 1.4b). These processes are most commonly employed in foundries. The design and process for sand casting is discussed in Chap. 7. Expendable-pattern expendable mold casting (EPEMC) processes involve temporary patterns with single-use mold; the examples include investment casting, lost foam casting, and full mold casting (see Fig. 1.4c). EPEMC processes are explained in Chap. 9.

　　Permanent mold casting processes are based on permanent, multiple-use molds; these processes can be divided into two groups: die casting and centrifugal casting (see Fig. 1.4d). Die casting processes can further be divided into the following type of processes: (a) basic permanent-mold casting or gravity die casting (GDC), (b) low-pressure die casting, and (c) high-pressure die casting, HPDC (see Chap. 10). Centrifugal casting processes include true centrifugal casting, semi-centrifugal casting, and centrifuge casting (see Chap. 11).

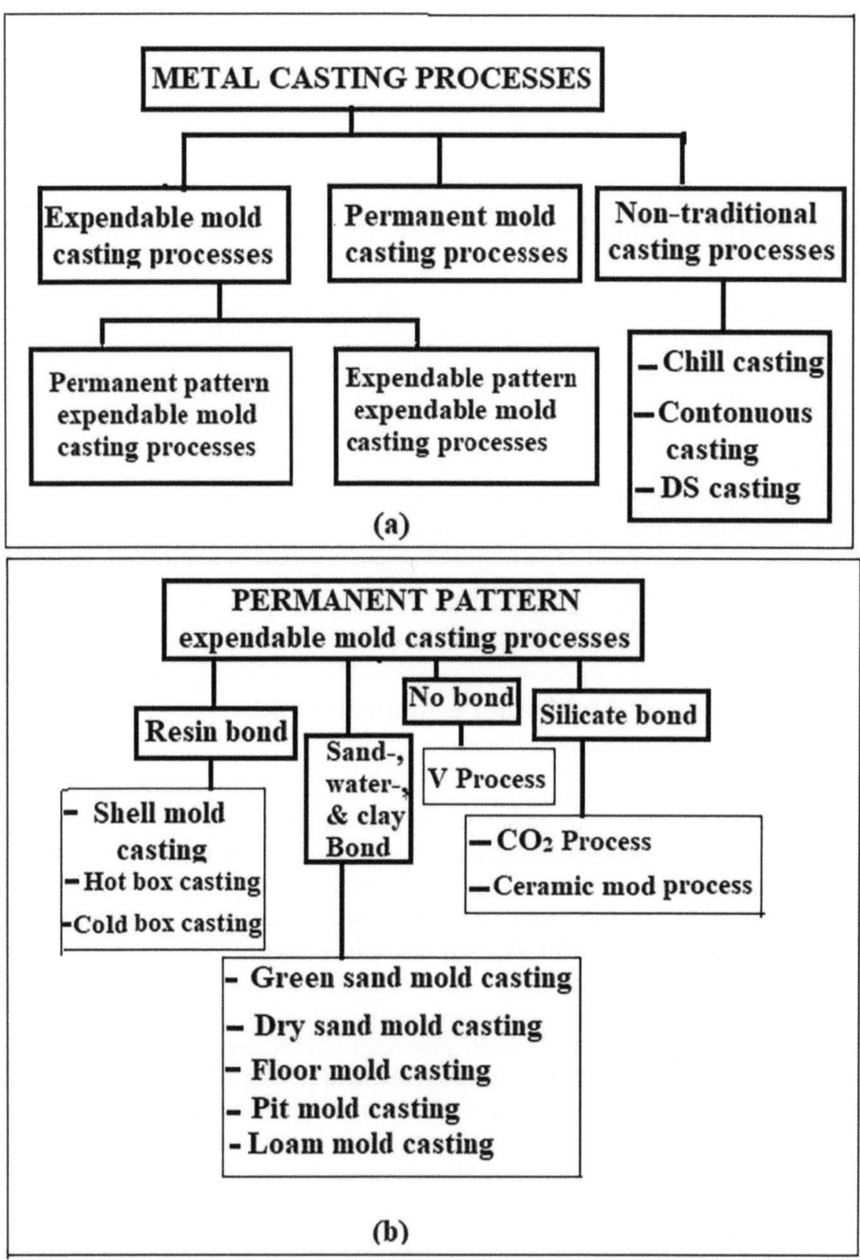

Fig. 1.4 Classification of metal casting processes: (**a**) primary classification, (**b**) permanent-pattern expendable-mold casting processes, (**c**) expendable-pattern expendable-mold casting processes, and (**d**) permanent mold casting processes

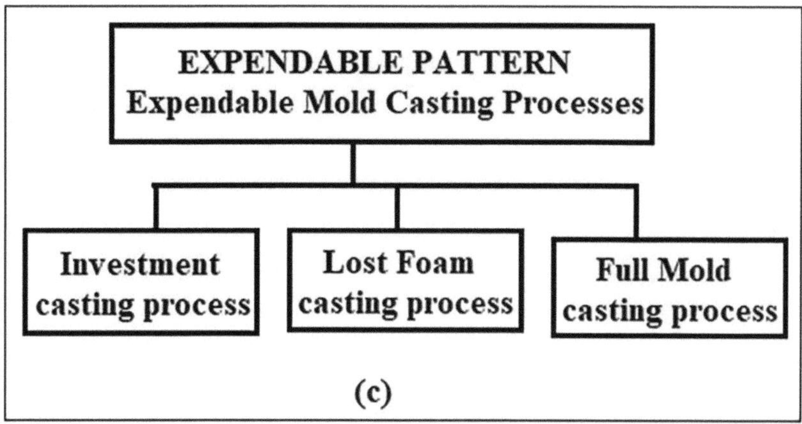

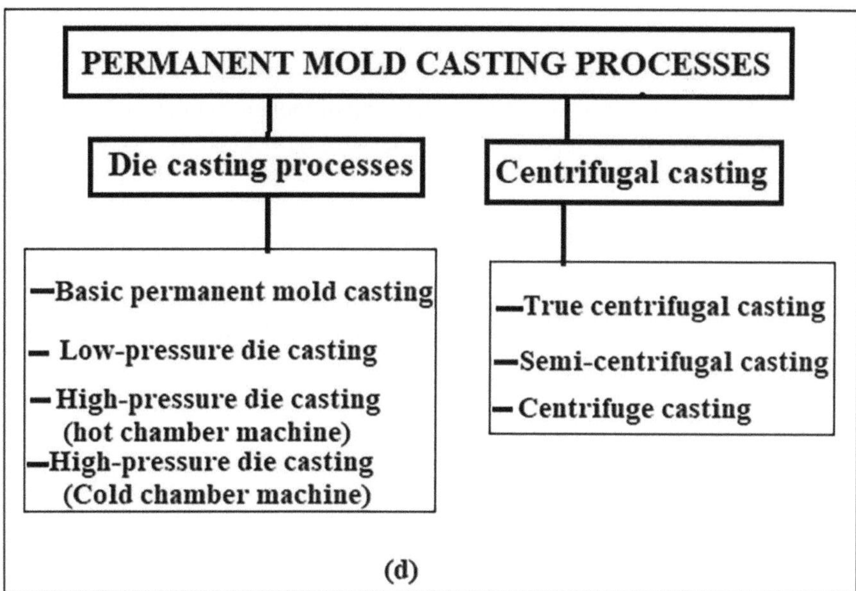

Fig. 1.4 (continued)

1.5   Non-traditional Casting Processes

1.5.1   Chill Casting Process

Chill casting is so named because it involves the use of a *chill*, which is an object that speeds up solidification in a specific portion of a metal-casting mold. In general, a chill is a metal object, but it can be made of any material that has a higher specific heat and/or heat conductibility as compared to the molding material. The use of

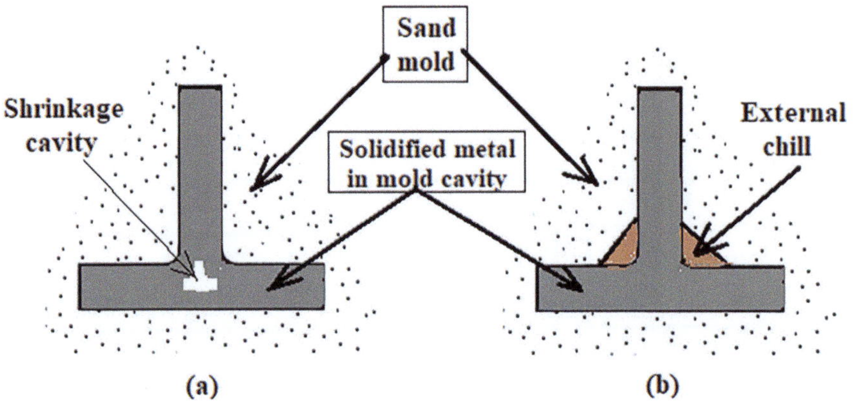

Fig. 1.5 Chill casting: (**a**) mold without chill showing shrinkage cavity, (**b**) external chills showing no shrinkage cavity defect

chills, in metal casting, encourages rapid freezing of molten metal, thereby preventing the occurrence of casting defects. There are two types of chills used in the casting process: (a) internal chills and (b) external chills. Internal chills are placed in the mold cavity. When molten metal comes in contact with the solid internal chill, it starts melting resulting in progressive directional solidification in the casting process. External chills are placed on the edge of the molding cavity, and effectively become part of the wall of the mold cavity. Chills have the best application where the thickness of the casting is large or uneven, since they prevent shrinkage-cavity and porosity defects (see Fig. 1.5).

1.5.2 Continuous Casting Process

Continuous casting of steel was first developed in the 1950s; it is now a standard manufacturing method (see Fig. 1.6). Continuous casting (CC) is applicable to steel as well as other metals and alloys. Continuous casting process involves the following major steps: (a) mold preparation, (b) melting of metal, and (c) metal forming. These major steps are explained as follows.

Mold Preparation The selection of mold material is very important in continuous casting. The mold materials must have excellent thermal conductivity and oxidation resistance. This is why copper and graphite are the best choices for mold making. The mold is machined into the required size and shape; it is cooled by circulating water during the CC process.

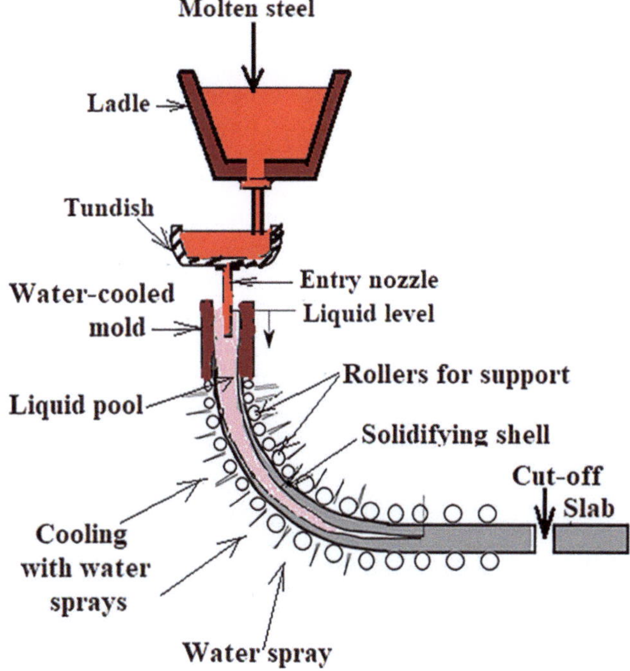

Fig. 1.6 Schematic of continuous casting of steel

Melting of Metal The metal is melted in a furnace; it is important to raise the temperature of the metal reasonably above its melting point to account for *superheat* (see Chap. 3, Sect. 3.3). For example, the metal should be heated to slightly above 700 °C for aluminum, since its melting point is around 660 °C.

Metal Casting/Forming The molten metal is then poured out of the ladle into an intermediate container—*tundish* (see Fig. 1.6). The metal from the *tundish* flows through a water-cooled mold; here the metal solidifies around the periphery of the mold due to a *chilling* effect. The resulting strand, which is still liquid on the inside, is continuously drawn downwards out of the mold. The strand, in an arch-shaped cooling chamber, is deflected and sprayed with water before it is finally straightened and cut-off to length to produce a slab. There are 2-strand and 4-strand CC machines, in use, in industrial practice.

There are a number of advantages of the CC process. Continuous casting results in improved product quality owing to a homogeneous and dense structure. Continuous casting enables us to produce long slabs, blooms, and tubes. Additionally, manufacturing cost is lower due to less material wastage and reduced energy consumption. However, there are a few limitations of CC process; the main disadvantage is the high set-up cost. The CC process is limited to more simple shapes that have a stable cross-section.

1.5.3   Directional Solidification Casting Technology

Directional solidification (DS) is a casting technique whereby the temperature of the mold is precisely controlled to promote the formation of aligned stiff crystals as the molten metal cools. The DS casting technology is widely used to manufacture superalloy columnar-grained (CG) and single-crystal (SC) blades for application in gas-turbine engines. DS casting process, for manufacturing superalloy turbine-blades, is accomplished in vacuum by pouring molten superalloy into a ceramic mold that has been pre-heated to a temperature above the liquidus of the alloy (Dull, 1987). The mold is open at the bottom and sits on a water-cooled copper chill. The alloy, upon contact with the chill, solidifies by nucleation and growth of grains that are most closely aligned with the [001] direction, thereby producing a columnar-grained blade. A single-crystal (SC) casting is obtained by including a helical con-struction (selector) above the starter block; the helical selector acts as a filter, and only permits a single grain to pass through (see Fig. 1.7). The SC turbine blades have excellent resistance to high-temperature creep (Huda, 2020).

Recently (2019), Liu and co-researchers have reported the development of the DS casting technology with the introduction of the liquid metal cooling (LMC) process. They have shown that the holding time after pouring (3–5 min) can pro-mote the growth of competitive grains and avoid a great deviation of columnar grains along the crystal orientation [001], resulting in a straight and uniform grain structure. In addition, in order to avoid the formation of wrinkles and to ensure a smooth blade surface, the withdrawal rate should be no greater than the growth rate of grain (Liu et al., 2019).

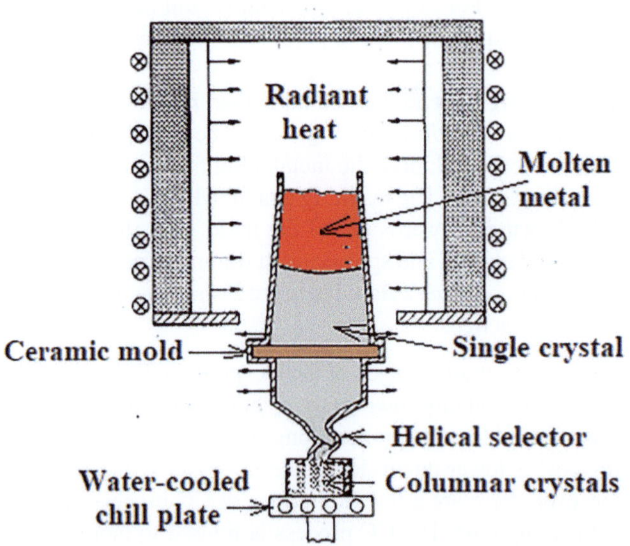

Fig. 1.7 DS casting process for manufacturing SC turbine blade

Very recently (2024), Zhang and co-workers have studied the effect of withdrawal rate on freckle formation of large-size directionally solidified (DS) nickel-based superalloy blades. They have experimentally proved that low and middle withdrawal rates are favorable to prevent the formation of freckle defect (Zhang et al. 2024).

Questions

1.1. Draw a labeled diagram showing the sand casting process.

1.2. (a) List and explain the principal steps in the sand casting process.

(b) Why is it important to properly design and perform each step of a casting process?

1.3. (a) What are the advantages of metal casting process over other manufacturing processes?

(b) Highlight the industrial applications of metal-cast products.

1.4. (a) Explain chill casting continuous casting processes.

(b) Define the term directional solidification (DS).

MCQs: 1.5. Encircle the most appropriate answers for each of the following statements.

(1) The metallic object used to promote solidification in a specific portion of a mold, is called:

(a) Riser, (b) sprue, (c) chill, (d) runner, (e) core.

(2) For which material was continuous casting process developed?

(a) Steel, (b) aluminum, (c) cast iron, (d) zinc, (e) copper.

(3) Sand casting process is an example of ___________:

(a) Permanent-mold casting, (b) expendable-pattern expendable-mold casting, (c) non-traditional casting, (d) permanent-pattern expendable-mold casting.

(4) Which casting technology is the most economical and appropriate for castings with large and uneven thicknesses?

(a) Chill casting, (b) centrifugal casting, (c) continuous casting, (d) investment casting.

(5) Which casting technology is used to manufacture single-crystal superalloy turbine-blades?

(a) Die casting, (b) fast solidification, (c) directional solidification, (d) slow solidification.

(6) Lost foam casting process is an example of ____?

(a) Permanent-mold casting, (b) expendable-pattern expendable-mold casting, (c) permanent-pattern expendable-mold casting, (d) high-tech casting process.

References

Dull, D. N. (1987). Directionally solidified superalloys. In C. T. Sims, N. S. Stoloff, & W. C. Hagel (Eds.), *Superalloys II*. Wiley.

Huda, Z. (2020). *Metallurgy for physicists and engineers*. CRC Press.

Liu, X.-f., Lou, Y.-c., Su, G.-q., et al. (2019). Directional solidification casting technology of heavy-duty gas turbine blade with liquid metal cooling (LMC) process. *Research & Development, China Foundry, 16*, 23–30. Quaker City Castings, 2024, https://quakercitycastings.com/#

Zhang, Y., Jia, Y., & Zhoa, J. (2024). Effect of withdrawal rate on freckle formation of large-size directional solidified nickel-based superalloy blades. *Journal of Materials Research and Technology, 30*, 1518–1530.

Part I
Melting and Solidification Design

Chapter 2
Melting and Flow of Metals

Nomenclature

Q	Heat energy, J
H_f	Latent heat of fusion, J
Q	Volume flow rate, m^3/s
Cs	Specific heat of solid metal, J/kg•°C
CL	Specific heat of liquid metal, J/kg•°C
T_o	Starting or ambient temperature
T_m	Melting temperature
T_p	Pouring temperature
E	Energy
ρ	Density, kg/m^3
P	Pressure, Pa
h	Head or height or sprue length, m
V	Volume, m^3
v	velocity, m/s
A	Area of cross-section, m^2
t_F	Time to fill mold cavity, s
t	Pouring time, s
R	Pouring rate, kg/s
R_{eff}	Effective pouring rate, kg/s
W	Mass, kg

2.1 Heat Energy Analysis for Melting and Pouring

It is mentioned in the preceding chapter that melting, for casting, requires super-heating, i.e., heating the metal to the pouring temperature that is reasonably higher than the melting point of the metal. The total heat energy required for melting the metal (for casting) can be expressed by the following relationship (Huda, 2018):

Z. Huda, *Metal Casting Engineering*, Mechanical Engineering Series,
https://doi.org/10.1007/978-3-031-84620-5_2

17

$$Q = Q_{\mathrm{I}} + Q_{\mathrm{II}} + Q_{\mathrm{III}} \tag{2.1}$$

where Q is total heat energy required to heat the metal from the starting (ambient) temperature to a pouring temperature, Q_I is heat energy to heat the metal to raise its temperature from starting temperature to the melting point, Q_{II} is the heat energy required to melt the metal, and Q_{III} is heat energy required in superheating, i.e., to raise temperature of molten metal from melting point to the pouring temperature (see Fig. 2.1).

The heat energy Q_I (in J) can be calculated by:

$$Q_{\mathrm{I}} = m\,C_s \cdot \left(T_m - T_o\right) \tag{2.2}$$

where m is the mass of the metal being heated (kg), C_s is the specific heat of the solid metal (J/kg•°C), T_m is the melting temperature of the metal (°C), and T_0 is the starting temperature. In the case of casting an alloy, the melting temperature can be determined with reference to the relevant phase diagram, and then using Eq. 2.2, as illustrated in Examples 2.1 and 2.2. The heat energy Q_{II} (in J) can be determined by:

$$Q_{\mathrm{II}} = m\,H_f \tag{2.3}$$

where H_f is the heat of fusion of the metal (J/kg). The heat energy, Q_{III}, can be computed by:

$$Q_{\mathrm{III}} = m\,C_L \cdot \left(T_p - T_m\right) \tag{2.4}$$

where C_L is the specific heat of liquid metal (J/kg• °C), and T_p is the pouring temperature (°C).

By the combination of Eq. 2.1 and Eqs. 2.2, 2.3 and 2.4, we obtain:

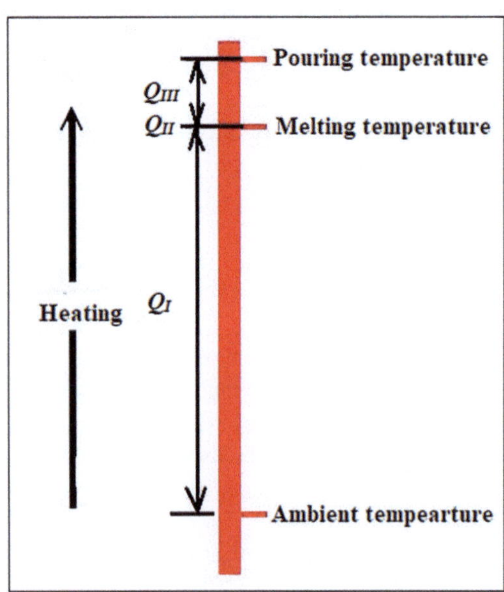

Fig. 2.1 Heat energies in melting and superheating of a metal

$$Q = m\,C_s \cdot \left(T_m - T_o\right) + m\,H_f + m\ C_L \cdot \left(T_p - T_m\right) \qquad (2.5)$$

Or

$$Q = m\left[C_s \cdot \left(T_m - T_o\right) + H_f + C_L \cdot \left(T_p - T_m\right)\right] \qquad (2.6)$$

The use of Eq. 2.6 requires reference to the thermal properties data for commonly used metals, which is presented in Table 2.1. The significance of Eq. 2.6 is illustrated in Example 2.3. Heating of metals, for casting, is accomplished by use of melting furnaces (see Chaps. 4 and 5).

2.2 Fluidity of Molten Metals

2.2.1 What Is Fluidity?

Fluidity is defined as the flow distance of molten metal in a channel of a uniform cross-section before the cessation of the flow caused by solidification (Chen & Li, 2024). The fluidity of a molten metal, in casting, measures the ease with which the

Table 2.1 Some thermal properties data of commonly used metals

Metal	Melting point (°C)	Specific heat of solid metal (J/kg• °C)	Specific heat of liq. Metal (J/kg• °C)	Heat of fusion (J/kg)
Aluminum	660	900	1130	4.00×10^5
Chromium	1857	460	940	3.94×10^5
Cobalt	1495	420	690	2.75×10^5
Copper	1085	390	490	2.06×10^5
Gold	1064	130	150	0.63×10^5
Iron	1540	450	820	2.47×10^5
Gray cast iron	1150–1250	$\cong$660	$\cong$950	$\cong 2.4 \times 10^5$
Lead	327	130	140	0.23×10^5
Magnesium	649	1050	1340	3.58×10^5
Manganese	1244	480	840	2.40×10^5
Molybdenum	2617	250	370	3.75×10^5
Nickel	1455	440	730	2.93×10^5
Platinum	1772	130	180	1.03×10^5
Silver	962	230	280	1.05×10^5
Sodium	98	1210	0.00	1.13×10^5
Tin	232	210	240	0.59×10^5
Titanium	1668	540	790	3.90×10^5
Tungsten	3410	130	170	1.90×10^5
Vanadium	1890	390	870	4.48×10^5
Zinc	420	390	–	1.12×10^5

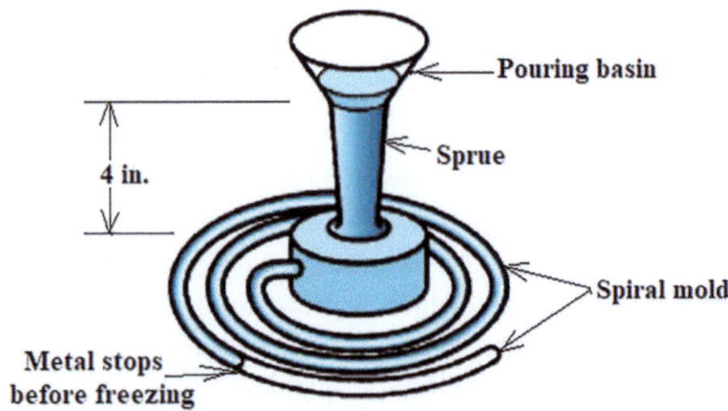

Fig. 2.2 Fluidity test using a spiral mold

molten metal can fill a mold cavity. It depends upon the freezing range of the molten alloy ($\Delta T = T_{\text{liquidus}} - T_{\text{solidus}}$), and the degree of superheat ($T_{\text{pouring}} - T_{\text{liquidus}}$) of the metal when it enters the mold. The fluidity of a poured metal is commonly determined by casting a fluidity spiral tube, which is a 60-in. long spiral that is molded into sand into which the metal is poured with a constant head of 4 in. (see Fig. 2.2). The length to which the metal penetrates the spiral measures the fluidity of the molten metal (Heine et al., 1976). It has been experimentally proved that alloys with large freezing ranges have low fluidity. On the other hand, alloys with a high degree of superheat have high fluidity.

2.2.2 Engineering Analysis of Fluidity

A successful casting practice requires good fluidity and high thermal conductivity of the metal. For example, gray cast iron has been shown to be one of the most fluid casting alloys with high thermal conductivity. It is therefore important that a foundry engineer first ascertains which type of cast iron (CI) s/he is going to melt: gray CI or white CI; this can be done by calculating the carbon equivalent (CE) for the CI. The CE of cast iron can be computed by (Huda, 2020):

$$\text{CE}\left(\text{in wt\%}\right) = \text{C} + \frac{Si + P}{3} \tag{2.7}$$

where C is the wt% carbon, Si is the wt% silicon, and P is the wt% phosphorous in the CI. A CE value close to the eutectic composition (CE $\geq$ 4.3) refers to gray CI (see Examples 2.4 and 2.5). On the other hand, CE < 4.3 indicates white cast iron.

Empirical fluidity spiral-test data for gray cast iron yields the following expression:

$$\text{Fluidity}\left(\text{in inches}\right) = 14.9 \times \text{CF} + 0.05 \cdot T_p - 155 \tag{2.8}$$

where CF is the composition factor of the gray iron, and T_p is the pouring temperature (in °F). The composition factor CF can be computed by:

$$CF = \%C + 0.25 \times \%Si + 0.5 \times \%P \tag{2.9}$$

The significance of Eqs. 2.8 and 2.9 is illustrated in Example 2.6. The fluidity values have been determined to be in the range of 30.8–41.30 in. The lowest fluidity value corresponds to a hypo-eutectic CF value at lower pouring temperatures, whereas the highest fluidity corresponds to CF values close to the eutectic at high pouring temperatures.

2.3 Fluid Flow: Bernoulli's Law

2.3.1 Fluid Flow System in a Casting Mold

The control of pouring is important in metal casting. It requires the analysis of fluid (molten metal) flow, which can be accomplished by the application of *Bernoulli's law*. According to *Bernoulli's law,* the sum of energies (head, kinetic, pressure, and friction) remains the same along a line of flow; it means that at any two points along a line of flow, the following relationship holds good:

$$E_1 = E_2 \tag{2.10}$$

where E_1 and E_2 are the sum of energies at points 1 and 2, respectively, along a line of flow (see Fig. 2.3) (see also Fig. 1.1).

The sum of energies E, along a line of flow in a fluid, can be expressed by (Flinn, 1963):

$$E = h + \frac{v^2}{2g} + \frac{P}{\rho} + z \tag{2.11}$$

where h = potential energy or head (cm), $\frac{v^2}{2g}$ is the kinetic energy, v is the flow velocity (cm/s), g is the gravitational acceleration ($= 981$ cm/s^2), P is the pressure on

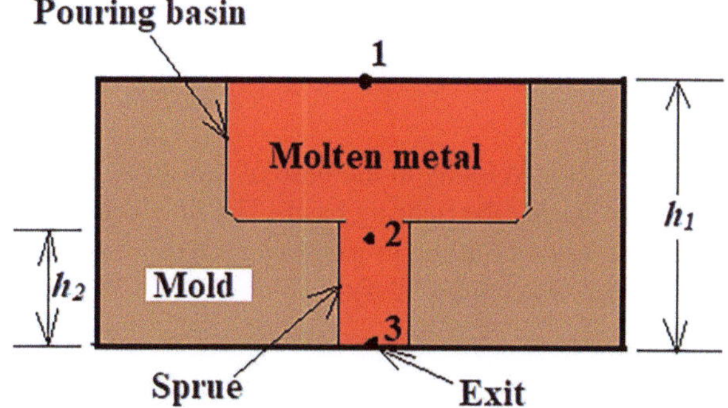

Fig. 2.3 Fluid flow system in a casting mold

the liquid (or molten metal) (N/cm^2), ρ is the density of the liquid (g/cm^3), and z is the catch-all term containing energy losses due to friction, turbulence, heat loss, etc. The velocity v and pressure P are critical dependent variables to control in gating systems with sand molds. This control enables us to quickly and cleanly fill the mold cavity. By combining Eqs. 2.10 and 2.11, we get:

$$h_1 + \frac{v_1^2}{2g} + \frac{P_1}{\rho} = h_2 + \frac{v_2^2}{2g} + \frac{P_2}{\rho} \tag{2.12}$$

where the terms in Eq. 2.12 have their usual meanings.

2.3.2   *Application of Bernoulli's Law for Flow Velocity*

It is useful to derive an expression for flow velocity of molten metal in the casting mold. Now, we refer to the molten metal flow in the gating system (Fig. 2.3). By using Bernoulli's law at points 1 and 3, we obtain:

$$h_1 + \frac{v_1^2}{2g} + \frac{P_1}{\rho} = h_3 + \frac{v_3^2}{2g} + \frac{P_3}{\rho} \tag{2.13}$$

It is assumed that the pouring basin remains full all the time; thus, $v_1 = 0$. Since the head is measured with reference to the exit point, $h_3 = 0$. Assuming that the system is in operation at normal condition of atmospheric pressure, we can write: $P_1 = P_3 = 1$ atm. Hence, Eq. 2.13 reduces, for point 1 and 3, to:

$$h_1 = \frac{v_3^2}{2g} \tag{2.14}$$

Or

$$h = \frac{v^2}{2g} \tag{2.15}$$

Or

$$v = \sqrt{2gh} \tag{2.16}$$

where v is the velocity at the base of the sprue, and h is the height or length of the sprue, including the pouring basin. The flow velocity at the base, $v_3 = v_b$, also represents the velocity at the in-gate, which requires the smallest cross-sectional area at the in-gate, A_b (see Fig. 2.4).

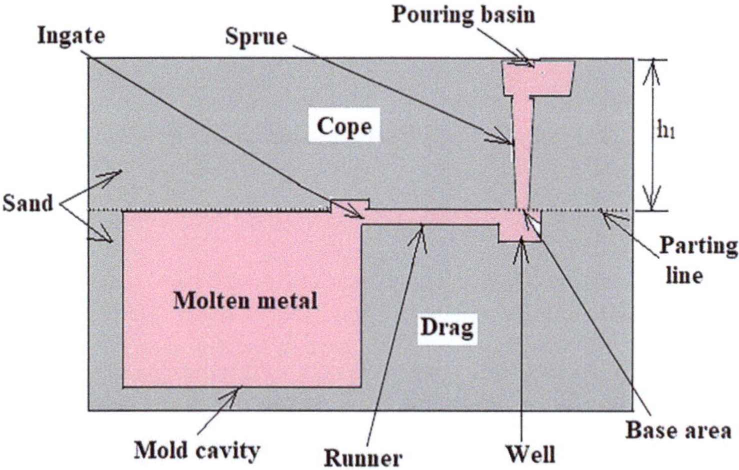

Fig. 2.4 A typical gating system showing a sprue, a well, a runner, and in-gate

The volume flow rate is equal to the flow velocity multiplied by the cross-sectional area of the fluid. The volume flow rate can be numerically, expressed as:

$$Q = v \cdot A = v_b \cdot A_b \tag{2.17}$$

where Q is the volume flow rate (m³/s), v_b is the flow velocity at the sprue base (m/s), and A_b is the cross-sectional area at the base (m²). The significance of Eqs. 2.16 and 2.17 is illustrated in Example 2.7.

The time to fill the mold cavity t_F (in s) can be computed by (Masoumi, 2007):

$$t_F = \frac{V}{Q} \tag{2.18}$$

where V is the volume of mold cavity (m³), and Q is the volume flow rate (m³/s) (see Example 2.8).

2.3.3   Application of Bernoulli's Law for Pressure

Bernoulli's law is also useful for quality control in metal casting, since it enables us to prevent aspiration defects. In order to avoid aspiration defect in a casting, the actual pressure within a flowing liquid, in a sand mold, must be kept above atmospheric pressure. In the case the pressure in the molten metal drops below 1 atmosphere, air can be aspirated (drawn in) into the metal stream, thereby causing defects within the casting. These defects may include gas porosity or the formation of metal oxides, a particularly serious problem in aluminum alloy castings which readily absorb hydrogen in the liquid state. The role of pressure and Bernoulli's law in casting defect is explained by the mathematical analysis presented in the following paragraph.

According to the law of conservation of mass, the mass remains constant along a line of flow. By applying the law of conservation of mass to the flowing molten metal, we can write:

$$A_2 \, v_2 = A_3 \, v_3 \qquad (2.19)$$

where A_2 and A_3 are the cross-sectional areas at points 2 and 3, respectively, and v_2 and v_0 are the fluid velocities at points 2 and 3, respectively (see Fig. 2.3). It is evident in Fig. 2.3 that the cross-sectional areas of the sprue at points 2 and 3 are assumed to be the same ($A_2 = A_3$). By using the law of continuity (Eq. 2.19), we get: $v_2 = v_3$.

The application of Bernoulli's Law at points 2 and 3 yields:

$$h_2 + \frac{v_2^2}{2g} + \frac{P_2}{\rho} = h_3 + \frac{v_3^2}{2g} + \frac{P_3}{\rho} \qquad (2.20)$$

Since $v_2 = v_3$ and $h_3 = 0$ (see Fig. 2.3.), Eq. 2.20 reduces to:

$$h_2 + \frac{P_2}{\rho} = \frac{P_3}{\rho} \qquad (2.21)$$

Or

$$P_2 = P_3 - \rho \cdot h_2 \qquad (2.22)$$

Since the system is assumed to be operating at atmospheric pressure, i.e., $P_3 = 1$ atm., the pressure at point 2 (P_2) is less than 1 atmosphere. This pressure condition is undesirable as gas aspiration defects will occur at this point in the system. It means that the use of a straight sprue (having the same cross-sectional area throughout) results in aspiration casting defects. This problem can be solved by correct sprue design, which is explained in Chap. 7 (Sect. 7.4). The significance of Eq. 2.22 is illustrated in Example 2.9.

2.4   Pouring Rate and Time in Casting

It is important for a foundry engineer to control the pouring of molten metal from the ladle into the mold, since control of pouring is a major factor in producing a good casting. A good pouring practice requires correct pouring temperature, uniform volume flow rate, and a minimum of turbulence, slag, entrapped sand, or other unwanted materials in the mold or molten metal system.

The pouring time basically depends on the type of metal being cast. Commonly used foundry alloys include gray/ductile cast iron, aluminum alloys, steels, copper alloys, and the like.

2.4.1   Pouring Rate and Time for Ferrous Alloys and Copper-Base Alloys

The optimum pouring rate for ferrous alloys and copper alloys can be determined by:

$$R = \frac{W^n}{1.34 + \dfrac{t}{13.77}} \tag{2.23}$$

where R is the pouring rate (kg/s), W is the mass of casting (kg), t is the critical casting thickness (mm), and n is a constant that depends on the casting mass (see Table 2.2).

The pouring time, for gray iron castings with thickness range of 2.5–15 mm and mass < 450 kg, can be expressed by:

$$t = S\sqrt{W} \tag{2.24}$$

where t is the pouring time (s), W is the mass of liquid metal poured (kg), and S is the coefficient taking into account the casting wall thickness (see Table 2.3). The significance of Eqs. 2.24 and 2.25 is illustrated in Examples 2.10 and 2.11.

2.4.2   Pouring Rate and Time for Aluminum-Silicon Casting Alloys

The pouring rate, for aluminum-silicon casting alloys, can be determined by (Jain, 2003):

$$R = b \bullet \sqrt{W} \tag{2.25}$$

Table 2.2 Constant n data for various ranges of casting mass for pouring rate

Casting mass, W (kg)	<500	500–5000	5000–15,000
Constant, n	0.50	0.67	0.70

Table 2.3 Coefficient S for various wall thicknesses of gray-iron casting

Casting wall thickness (mm)	2.5–3.5	3.5–8.0	8.0–15.0
S	1.63	1.85	2.20

Table 2.4 Coefficient b for various wall thicknesses of *Al-Si* casting

Casting wall thickness (mm)	<6	6–12	>12
Coefficient, b	1.63	1.85	2.20

where R is the pouring rate (kg/s), W is the mass of the casting (kg), and b is the coefficient for wall thickness for *Al-Si* alloys (see Table 2.4, see Example 2.12).

The effective pouring rate, for aluminum-silicon casting alloys, is given by:

$$ R_{eff.} = \frac{R}{KC} \tag{2.26} $$

where K is the metal fluidity (= 1 for *Al-Si* alloys), and C is the constant for the effect of friction for tapered sprues (C = 0.85–0.90).

The pouring time t can be determined by combining Eqs. 2.25 and 2.26, as follows:

$$ t = \frac{W}{R_{eff}} = \frac{\sqrt{W} \bullet K \bullet C}{b} \tag{2.27} $$

The significance of Eq. 2.27 is illustrated in Example 2.13.

2.5   Worked Numerical Examples in Melting and Flow of Metals

Example 2.1 Determining the T_m and the Heat Energy Required to Melt Al-Si Casting Alloy

Determine the heat energy required to raise the temperature of 5 kg *Al-15 wt% Si* casting alloy from the ambient temperature (20 °C) to its melting temperature. The specific heat of the solid alloy is 885 J/kg• °C. The phase diagram for aluminum-silicon alloy is shown in Fig. 2.5.

Solution

m = 5 kg, T_o = 20 °C, C_s = 885 J/kg• °C, Q_1 =?

In order to determine the melting temperature of *Al-15 wt% Si* alloy, we can draw a dotted vertical line in Fig. 2.5, to obtain the new Fig. 2.6.

Hence, the melting temperature of *Al-15 wt% Si* alloy is around 610 °C (see Fig. 2.6).

Now, the required heat energy can be computed using Eq. 2.2, as follows.

$$ Q_I = m\,C_S \cdot (T_m - T_o) = 5 \times 885 \times (610 - 20) $$
$$ = 2610750\,J = 2.611\ MJ $$

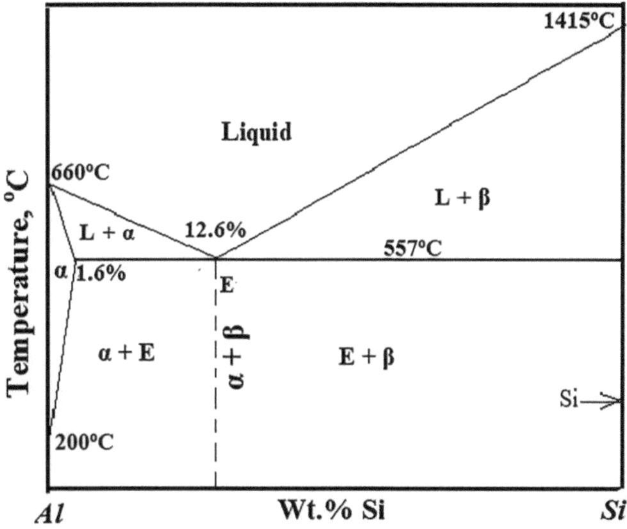

Fig. 2.5 Aluminum-silicon phase diagram

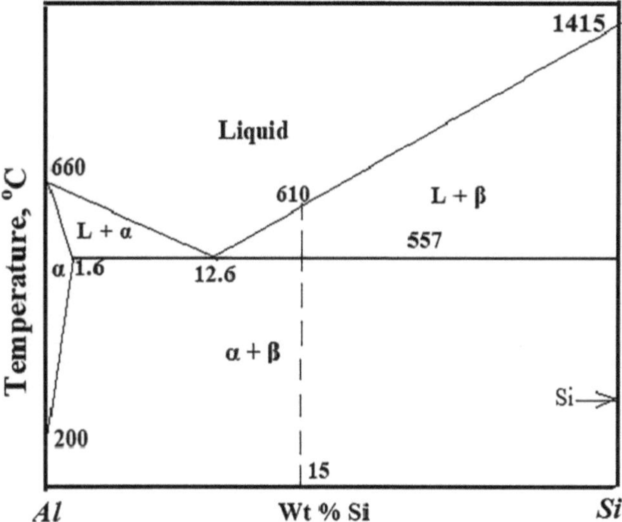

Fig. 2.6 *Al-Si* phase diagram showing the melting temperature of *Al-15%Si* alloy

Example 2.2 Determining the Heat Energy for Melting with the Aid of Microstructure

Assume you are an engineer in a foundry. The employee of your company pur-chased 4 kg aluminum-silicon alloy scarp with unknown composition (the scrap is made up of just one material). You conducted metallographic test for the scrap, which revealed a microstructure comprising of two-phase eutectic mixture: $\alpha + \beta$. Determine the heat energy required to raise the temperature of the alloy from the ambient temperature (20 °C) to its melting temperature. The specific heat of the solid alloy is 892 °C.

Solution

By reference to Fig. 2.5, the eutectic microstructure corresponds to the melting temperature of 557 °C for the alloy scrap. Hence, we can list the data for this problem as follows:

$m = 4$ kg, $T_o = 20$ °C, $T_m = 557$ °C, $C_s = 892$ J/kg• °C, $Q_1 = ?$
By using Eq. 2.2,

$$Q_1 = m\,C_s \cdot \left(T_m - T_o\right) = 4 \times 892 \times \left(557 - 20\right)$$
$$= 1916016\,J = 1.92 \ MJ$$

Example 2.3 Determining the Total Heat Energy Required for Melting the Metal for Casting

It is required to melt 8 kg commercially-pure solid nickel to produce a casting. The starting and the pouring temperatures are 20 °C and 1500 °C, respectively. Determine the total heat energy required to melt the metal for the casting process. Hint: Refer to the data in Table 2.1.

Solution

By reference to Table 2.1,

m = 8 kg, $T_0 = 20$ °C, $T_m = 1455$ °C, $T_p = 1500$ °C, $C_S = 440$ J/kg• °C, $C_L = 730$ J/kg• °C, $H_f = 293{,}000$ J/kg.

By using Eq. 2.6,

$$\mathbf{Q = m\left[\, C_s \cdot \left(T_m - T_o\right) + H_f + C_L \cdot \left(T_p - T_m\right)\right]}$$
$$= 8 \times \left\{\left[\,440 \times \left(1455 - 20\right)\right] + 293000 + \left[730 \times \left(1500 - 1455\right]\right\}$$

$$Q = 8 \times \left[\left(440 \times 1435\right) + 293000 + \left(730 \times 45\right)\right]$$
$$= 8\left(631400 + 293000 + 32850\right) = 7658000 \ J$$

The total heat energy required to melt the metal for the casting process $\cong 7.66$ MJ.

Example 2.4 Identifying Gray Cast Iron with the Aid of Carbon Equivalence

You are provided with two samples of cast irons (A and B). Sample A has the composition (in wt%): C: 3.4, Si: 1.6, P: 1.5, S: 0.1, Mn: 0.6, Fe: balance. The composition of sample B (in wt%) is: C: 2.4, Si: 1.4, P: 0.13, Mn: 0.4, Fe: balance. Identify which sample is gray cast iron and which one in white CI.

Solution

By using Eq. 2.7 for sample A,

$$CE\left(\text{in wt\%}\right) = C + \frac{Si + P}{3} = 3.4 + \frac{1.6 + 1.5}{3} = 4.43$$

By using Eq. 2.7 for sample B,

$$CE\left(\text{in wt\%}\right) = C + \frac{Si + P}{3} = 2.4 + \frac{1.4 + 0.13}{3} = 2.91$$

The CE value of sample A is close to 4.3, hence the sample A is gray CI. Sample B is white CI.

Example 2.5 Determining the Heat Energy for Casting with the Aid of CE of Cast Iron

Refer to the CE data for the gray cast iron in Example 2.4. It is required to raise the temperature of 7 kg of the cast iron from ambient temperature (20 °C) to the pouring temperature (1370 °C). Determine the heat energy required for the job. The Fe-Fe$_3$C phase diagram is shown in Fig. 2.7.

Solution

By reference to the CE value for the gray cast in Example 2.4, the carbon equivalence (CE) is 4.43 wt%. Next, we can use the CE value (= 4.43) to determine the

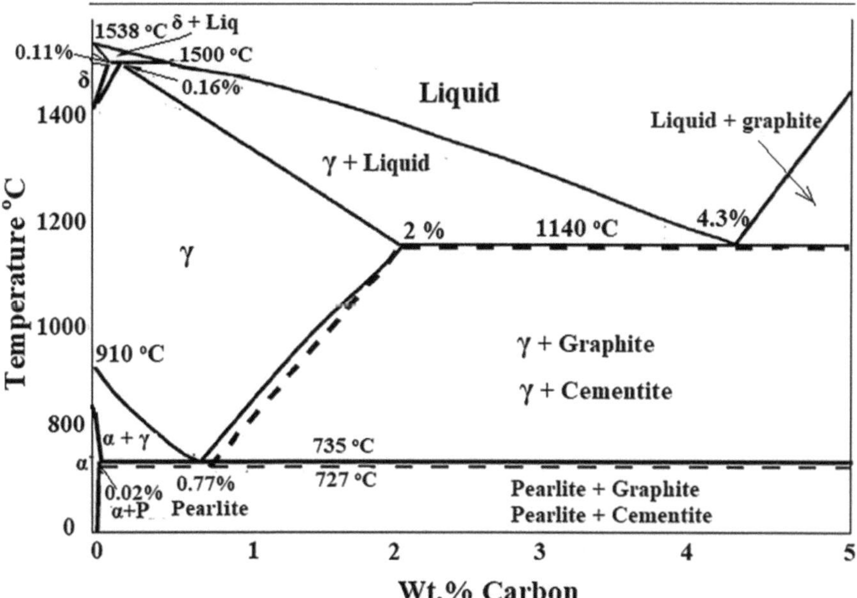

Fig. 2.7 The Fe-Fe$_3$C equilibrium diagram

melting temperature of the iron, as follows. By reference to the Fe-Fe$_3$C equilibrium diagram (Fig. 2.7), we get the corresponding melting temperature as: $T_m \cong$ 1180 °C. Now, by reference to the data in Table 2.1, the final data for this numerical example becomes:

m = 7 kg, T_o = 20 °C, T_m = 1180 °C, T_p = 1370 °C, C_S = 660 J/kg• °C, C_L = 950 J/kg• °C, H_f = 240,000 J/kg.

By using Eq. 2.6,

$$Q = m\left[C_s \cdot (T_m - T_o) + H_f + C_L \cdot (T_p - T_m) \right]$$
$$= 7 \times \left\{ \left[660 \times (1180 - 20) \right] + 240000 + [950 \times (1370 - 1180]\right\}$$

$$Q = 7 \times \left[(660 \times 1160) + 240000 + (950 \times 190) \right]$$
$$= 7 \times (765600 + 240000 + 180500) = 8302700 \ \text{J}$$

The total heat energy required = 8.3 MJ.

Example 2.6 Determining the Fluidity of Gray Cast Iron at Given Pouring Temperature

Determine the fluidity of gray cast iron that has the following composition (in wt%): 3.2 C, 1.8 Si, 1.4 P, 0.1 S, 0.6 Mn, and balance: Fe. The pouring temperature is 2550 °F.

Solution

By using Eqs. 2.8 and 2.9,

$$CF = \%C + 0.25 \times \%\text{Si} + 0.5 \times \%\text{P}$$
$$= 3.2 + (0.25 \times 1.8 + 0.50 \times 1.4) = 4.35$$

$$\text{Fluidity} = 14.9 \times CF + 0.05 \cdot T_p - 155$$
$$= (14.9 \times 4.35) + (0.05 \times 2550) - 155 = 37.31 \ \text{in.}$$

Example 2.7 Determining the Flow Velocity and Volume Flow Rate in a Gating System

The down-sprue, leading into the runner of a mold, has a length (including the pouring basin) of 170 mm. The cross-sectional area at the sprue-base is 360 mm^2. Determine the (a) flow velocity at the base, and (b) volume rate of flow.

Solution

h = 170 mm = 0.17 m, A_b = 360 mm^2 = 360 × 10^{-6} m^2, g = 9.81 m/s^2, v =?, Q =?

By using Eqs. 2.16 and 2.17,

$$v = \sqrt{2gh} = \sqrt{2 \times 9.81 \times 0.17} = 1.826\,\text{m}/\text{s} = v_b$$

$$Q = v_b \cdot A_b = 1.826 \frac{m}{s} \times 360 \times 10^{-6} \ m^2$$
$$= 657.471 \times 10^{-6} \ m^3/s = 657.471 \ cm^3/s$$

Example 2.8 Determining the Time to Fill the Mold Cavity

By using the data in Example 2.7, determine the time to fill the mold cavity, if dimensions of the casting are: 200 mm × 150 mm × 100 mm.

Solution

Volume of mold cavity = V = volume of casting = 200 mm × 150 mm × 100 mm = 3 × 10^6 mm^3

V = 3 ×10^6 × 10^{-9} m^3 = 3 ×10^{-3} m^3, Q = 657.471 × 10^{-6} m^3/s.

By using Eq. 2.18,

$$\text{Time to fill the mold cavity} = t_F = \frac{V}{Q}$$

$$= \frac{3 \times 10^{-3}}{657.471 \times 10^{-6}} = 4.563 \ s$$

Example 2.9 Determining the Pressure at a Point in a Casting Gating System

Consider the flow of gray cast iron in a casting gating system with straight-sided sprue. The head at a point P, along a line of flow, is 20 mm. The casting gating system is operating at normal atmospheric pressure, i.e., P_3 = 1 atm. The density of cast iron is 7250 kg/m^3. Calculate the pressure at the point P. Is the calculated pressure value desirable for defect-free casting?

Solution

h_2 = 20 mm = 0.02 m, ρ = 7250 kg/m^3, P_3 = 1 atm = 101,300 Pa, P_2 =?

By using Eq. 2.22,

$$P_2 = P_3 - \rho \cdot h_2$$
$$= 101.3 \times 10^3 - (7250 \times 0.02)$$
$$= 101155 \ Pa$$

Since the calculated pressure is less than 1 atm, it is *not* desirable for defect-free casting.

Example 2.10 Determination of Pouring Rate for a Copper Alloy Casting

A 10-kg brass plate with a critical thickness of 5.6 mm is to be cast. Determine the pouring rate for the job.

Solution

By reference to Table 2.2, n = 0.50, W = 10 kg, n = 0.50, t = 5.6 mm, R =?

By using Eq. 2.23,

$$\text{Pouring rate} = R = \frac{W^n}{1.34 + \dfrac{t}{13.77}} = \frac{10^{0.5}}{1.34 + \dfrac{5.6}{13.77}}$$

$$= \frac{3.162277}{1.34 + 0.407} = 1.81 \ \text{kg/s}$$

Example 2.11 Determination of Pouring Time for Gray Iron Casting
A 25-kg gray iron casting, with a wall thickness of 4.0 mm, is required to be produced. Determine the pouring time for the job.

Solution
By reference to Table 2.3, S = 1.85.
   By using Eq. 2.24,

$$\text{Pouring time} = t = S\sqrt{W} = 1.85 \times \sqrt{25} = 9.25\,\text{s}$$

Example 2.12 Determination of Pouring Rate for Aluminum-Silicon Casting
The pattern dimensions for a rectangular Al-Si casting are: 860 mm × 580 mm × 25 mm. The density of the alloy is 2.5 g/cm³. Determine the pouring rate for the casting job.

Solution
By reference to Table 2.4 for 10-mm thickness casting, b = 1.85;
   Density = ρ = 2.5 g/cm³ = 2.5 × 10³ kg/m³ = 2500 kg/m³, R =?
   Volume = V = 860 mm × 580 mm × 25 mm = 0.86 × 0.58 × 0.025 m³ = 0.01247 m³.

$$W = \rho \cdot V = 2500 \times 0.01247 = 31.175 \ \text{kg}$$

By using Eq. 2.25,

$$\text{Pouring rate} = R = b \cdot \sqrt{W}$$

$$= 2.20 \times \sqrt{31.175} = 12.28 \ \text{kg/s}$$

Example 2.13 Determination of Pouring Time for Aluminum-Silicon Casting
By using the data in Example 2.12, determine the pouring time for the casting job. Take the constant for the effect of friction for tapered sprue = 0.85.

Solution
W = 31.175 kg, b = 1.85, C = 0.85, K = 1, t =?
   By using Eq. 2.27,

$$\text{Pouring time} = t = \frac{\sqrt{W} \cdot K \cdot C}{b}$$

$$= \frac{\sqrt{31.175} \times 1 \times 0.85}{1.85} = 2.565 \ \text{s}$$

Example 2.14 Determining the Mass of Metal to be Melted Considering the Wastage

An aluminum component (with dimensions 30 cm × 22 cm × 10 cm) is required to be cast by sand casting. Assuming that there is 4% wastage during melting, determine the mass of aluminum scrap to be added to the melting furnace. Density of aluminum is 2.7 kg/cm³.

Solution

Component dimensions: 30 cm × 22 cm × 10 cm, $\rho = 2.7$ kg/cm³, Wastage = 4%.

$$\text{Volume of the component} = V = 30 \ \text{cm} \times 22 \ \text{cm} \times 10 \ \text{cm}$$
$$= 6600 \ \text{cm}^3$$

$$\text{Mass of the aluminum scrap} = \rho * V = 2.7 \times 6600$$
$$= 17820 \ \text{g} = 17.820 \ \text{kg}$$

$$\text{Mass of aluminum scrap to be melted} = 17.820 + 4\% \text{of} 17.820$$
$$= 17.820 + 0.713 = 18.53 \ \text{kg}$$

Mass of aluminum scrap to be added to the melting furnace = m = 18.53 kg.

Example 2.15 Determining the Total Heat Energy for Casting Taking into Account Wastage

By using the data in Example 2.14, determine the heat energy required to raise the temperature of the metal from the ambient temperature (25 °C) to the pouring temperature of 780 °C. Use the thermal properties data in Table 2.1.

Solution

m = 18.53 kg, $T_o = 25$ °C, $T_p = 780$ °C, $T_m = 660$ °C, $c_s = 900$ J/kg•°C, $c_L = 1130$ J/kg• °C, $H_f = 400{,}000$ J/kg, Q =?

By using Eq. 2.6,

$$Q = m\left[C_s \cdot (T_m - T_o) + H_f + C_L \cdot (T_p - T_m)\right]$$
$$- 18.53 \times \left\{\left[900 \times (660 - 25)\right] + 400000 \mid [1130 \times (780 \quad 660]\right\}$$

$$Q = 18.53 \times (571500 + 400000 + 135600)$$
$$= 1107100 \ \text{J} = 1.11 \ \text{MJ}$$

Questions and Problems

2.1. Encircle the most appropriate answers for each of the following statements:

   (1) The distance of metal flow before it solidifies and stops is called:
         (a) Flow velocity, (b) fluidity, (c) solidification, (d) distance of flow.
   (2) The fluidity of a poured metal is commonly determined by casting a
        __________.
         (a) Fluidity spiral tube, (b) fluidity straight tube, (c) fluidity spiral bar,
        (d) none.
   (3) Alloys with large freezing ranges have ____________.
         (a) High fluidity, (b) high freezing point, (c) low freezing point, (d) low
        fluidity.
   (4) The product of the flow velocity and the cross-sectional area of the fluid is
        called___:
         (a) Velocity flow rate, (b) area flow rate, (c) volume flow rate. (d)
        fluid rate.
   (5) Aspiration casting defect is caused by __________:
         (a) Tapered sprue, (b) straight sprue, (c) spiral sprue, (d) square sprue
   (6) The ratio of the volume of mold cavity to the volume flow rate is equal to
        _________:
         (a) Time of solidification, (b) time to fill mold cavity, (c) flow rate
        of fluid
   (7) A distinction between gray cast iron and white cast iron can be made
        by______:
         (a) Carbon equivalent, (b) composition equivalence, (c) gray equivalence

2.2. (a) Define the term "fluidity".
     (b) Explain the fluidity test using a spiral mold with the aid of a sketch.
2.3. (a) Why is gray cast iron considered as one of the best alloys for casting?
     (b) Specify the range of fluidity values and the corresponding casting
conditions.
2.4. (a) Draw the sketch showing fluid flow system in a casting mold with a straight
down-sprue.
     (b) State Bernoulli's law and give its mathematical expression.
     (c) Apply Bernoulli's law to derive an expression for the flow velocity.
2.5. (a) State the law of conservation of mass, and give its mathematical expression
for flowing molten metal.
     (b) Apply Bernoulli's law to derive an expression for the pressure at a point
in a fluid flow system in a casting mold with a straight down-sprue.
     (c) What problem is encountered in a fluid flow system with a straight
sprue? And why?
     P-2.6. A gray cast iron has the following eutectic composition (wt%): 3.5 C,
2.4 Si, 0.04 P. Determine the fluidity of the iron at a pouring temperature of
2700 °F.
     P-2.7. Determine the heat energy required to raise the temperature of 5 kg
Al-20 wt% Si casting alloy from the ambient temperature (20 °C) to its melting

point. The specific heat of the solid alloy is 832 J/kg• °C. The phase diagram for *Al-Si* system is shown in Fig. 2.5.

P-2.8. It is required to melt 400 g commercially-pure solid gold to produce a casting. The starting and the pouring temperatures are 20 °C and 445 °C, respectively. Determine the total heat energy required to melt the metal for the casting process. Refer to the data in Table 2.1.

2.9. A cast iron has the following composition (in wt%): C: 3.2, Si: 1.8, P: 1.4, S: 0.1, Mn: 0.6, Fe: balance. It is required to raise the temperature of 8 kg of the cast iron from ambient temperature (20 °C) to the pouring temperature (1350 °C). Determine the heat energy required for the job. The $Fe-Fe_3C$ equilibrium diagram is shown in Fig. 2.7.

2.10. The down-sprue, leading into the runner of a mold, has a length of 150 mm. The cross-sectional area at the sprue-base is 380 mm². Determine the (a) flow velocity at the base, and (b) volume rate of flow.

2.11. A 8-kg bronze plate with a critical thickness of 4.5 mm is to be cast. Determine the pouring rate for the job.

2.12. A 20-kg gray iron casting, with a wall thickness of 3.0 mm, is required to be produced. Determine the pouring time for the job.

References

Chen, G.-C., & Li, X. (2024). Effect of TiC nano-treating on the fluidity and solidification behavior of aluminum alloy 6063. *Journal of Materials Processing Technology, 324*, 118241.

Flinn, R. A. (1963). *Fundamentals of metal casting*. Addison-Wesley Publishers.

Heine, R. W., Loper, C. R., & Rosenthal, P. C. (1976). *Principles of metal casting* (2nd ed.). McGraw Hill.

Huda, Z. (2018). *Manufacturing—Mathematical models, problems, solutions*. CRC Press.

Huda, Z. (2020). *Metallurgy for physicists and engineers*. CRC Press.

Jain, P. L. (2003). *Principles of foundry technology* (4th ed.). McGraw-Hill Company Limited.

Masoumi, A. (2007). Effect of gating design on mould filling. In *International journal of cast metal research*. American Foundry Society.

Chapter 3
Solidification Design

Nomenclature

N	Nucleation rate
G	Growth rate
C_{seg}	Elemental micro-segregation level
C_{max}	Maximum elemental solid solubility at the end of solidification
C_{min}	Minimum solid solubility in the initial stage of solidification
C_{ave}	Average elemental content (solid solubility)
T_L	Liquidus temperature
T_S	Solidus temperature
ΔT	Freezing range
T_p	Pouring temperature
T_m	Melting temperature
$T_p - T_m$	Superheat
$\dfrac{\partial T}{\partial t}$	Rate of heat transfer
x	Distance from metal-mold interface
α_M	Thermal diffusivity of mold
K_M	Thermal conductivity of the mold
ρ_M	Density of the mold
C_M	Specific heat of the mold material at constant pressure
t_s	Total solidification time
ΔH_f	Heat of fusion
V_{Metal}	Volume of metallic casting
A	Surface area of the casting through which heat is extracted
C_m	Mold constant
n	Constant in *Chvorinov's* equation
ε	Shape coefficient in the modified *Chvorinov's* equation
$\dot{\theta}$	Cooling rate
d	Grain size

Z. Huda, *Metal Casting Engineering*, Mechanical Engineering Series,
https://doi.org/10.1007/978-3-031-84620-5_3

3.1   Solidification and Microstructural Evolution

It has been established in previous chapters that metal casting involves a great deal of pouring of liquid metal followed by its solidification in the mold cavity. The understanding of the mechanisms of solidification and the evolution of grained microstructure is of great technological importance, since some mechanical properties of the solidified poly-crystalline metal strongly depend on the rate of solidification/cooling. In general, the solidification of metal is accompanied by the nucleation and growth mechanism, which is explained as follows. At the beginning of solidification, multiple nuclei form, which grow in stages to complete recrystallization, thereby forming a grained microstructure. The solidification of a poly-crystalline (consisting of multiple crystals) metal occurs in the following four stages: (I) *nucleation*: multiple nuclei begin to form in the melt (Fig. 3.1a), (II) *crystal growth:* the dendritic growth of nuclei into crystals (Fig. 3.1b), (III) *formation of randomly oriented crystals* (Fig. 3.1c), and (IV) completion of solidification: formation of grained microstructure (Fig. 3.1d).

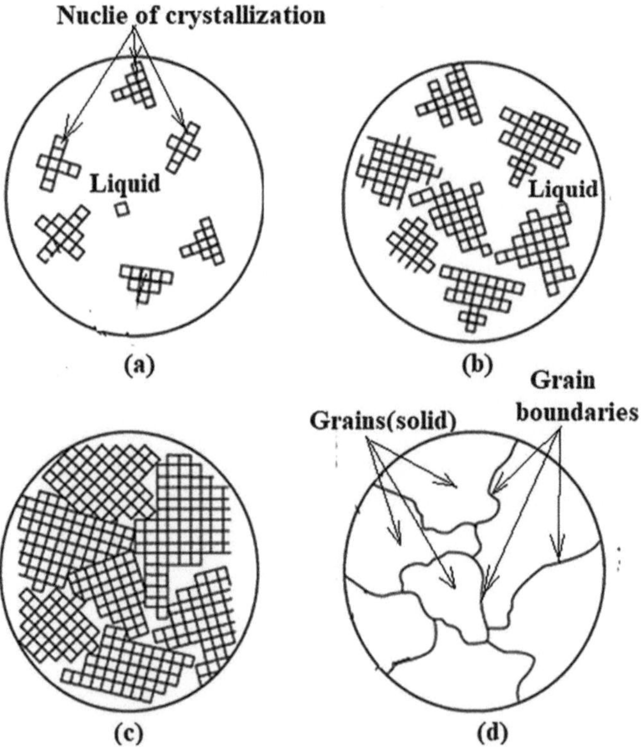

Fig. 3.1 Steps in crystallization during solidification of a pure metal: (**a**) nucleation in liquid, (**b**) growth of nuclei, (**c**) progress of crystallization, (**d**) grained microstructure—completion of solidification

Nucleation refers to the beginning of crystallization; at this stage multiple stable nuclei begin to grow at the nucleation points (see Fig. 3.1a). As the molten metal cools at the solidification temperature, atoms of the molten metal begin to bond together at the nucleation points, thereby forming tiny crystals. The *crystal growth* stage involves the dendritic growth of crystals (Fig. 3.1b). Then the grown crystals impinge upon adjacent growing crystals (Fig. 3.1c).

Finally, solidification is completed, resulting in a grained microstructure with grain boundaries. (Fig. 3.1d). A grain boundary is an array of dislocations, which are high-energy sites in the microstructure.

3.2   Rate of Solidification and Casting's Characteristics

3.2.1   *Rate of Solidification, Grain Size, and Casting's Strength*

It is learnt in the preceding section that solidification of molten metal is associated with a mechanism that involves nucleation and growth. The sizes of the individual crystals (or grains) depend on the number of nucleation points (or the rate of nucleation, N) and the rate of growth of the nuclei, G. The nucleation rate (N) refers to the number of stable nuclei formed per unit time, whereas the rate of growth (G) refers to the average increase in the nucleus (crystal) size per unit time. In the case the rate of solidification is high (fast cooling), the rate of nucleation exceeds the rate of growth of the nuclei (N > G); this mechanism results in a fine-grained microstructure (Fig. 3.2a). On the other hand, if the rate of solidification is low (slow cooling), the rate of crystal growth exceeds the nucleation rate (N < G), thereby resulting in a coarse grain-sized material (Fig. 3.2b). The effect of cooling rate on the grain size is quantitatively illustrated in Examples 3.1, 3.2, 3.3, 3.4 and 3.5.

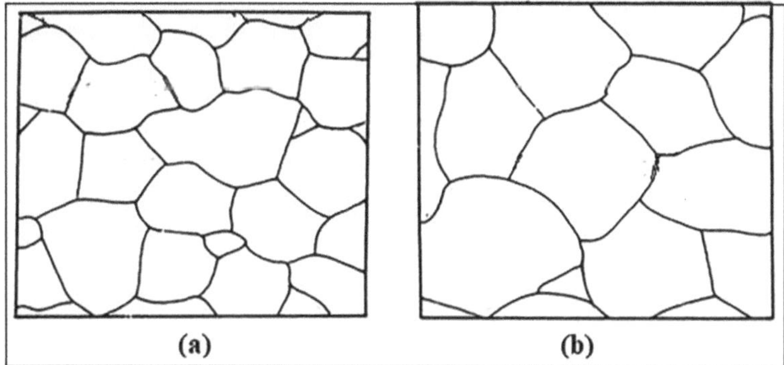

Fig. 3.2 Schematic of grained microstructures of a cast metal: (**a**) fine-grained microstructure resulting from fast cooling, (**b**) coarse-grained microstructure resulting from slow cooling

The number of grains per unit area may be mathematically related to the rate of nucleation and the growth rate by (Cui & Qin, 2007):

$$Z_S = 1.1\sqrt{\frac{N}{G}} \tag{3.1}$$

where Z_S is the number of grains per unit area, N is the rate of nucleation, and G is the rate of growth. It is well established that a fine-grained microstructure indicates higher yield strength of the material, which is governed by Hall–Petch relationship (Hansen, 2004). The increase in yield strength with a decrease in grain size can be justified on the basis of grain-boundary strengthening (Huda, 2020). It is, therefore, concluded that a high rate of solidification (rapid cooling) results in a fine grain size, which in turn imparts higher yield strength and ultimate tensile strength in the casting. This scientific principle is applied in die casting for producing high-strength machine components (Andresen, 2004).

Recently (2022), Huang and co-researchers have studied the effects of cooling rate on the grain morphology and element segregation behavior of *Fe-Mn-Al-C* low-density steel (composition in wt%: 58.7Fe-30Mn-10Al-1.1C, approx.) during solidification. They have reported that the grain size of the austenitic phase gradually decreases from 120.17 μm to 88.63 μm as the cooling rate increases from 1.69 °C/s to 10.28 °C/s (Huang et al., 2022).

3.2.2 Rate of Solidification and Micro-segregation

Prior to discussing micro-segregation, it is appropriate to define the term solid solution. A solid solution is a homogeneous mixture of two components in solid state and having a single crystal structure in the alloy. As a rule of thumb, higher the solid solubility, lesser will be the segregation in an alloy. *Micro-segregation* refers to the pattern of composition variation that remains in a solidified alloy. It has been experimentally proved that with the increasing cooling rate, solidified structures get significantly refined and the solid solubility of elements gets enhanced, reducing the content of second-phase particles and elemental micro-segregation (Liu et al., 2020; He et al., 2020; Tang et al., 2019). However, an exception to this solidification behavior has been reported with reference to the cooling rates and the resulting micro-segregation particularly in aluminum-zinc alloys, which is explained in the following paragraph.

The elemental micro-segregation can be related to the maximum and minimum solid solubilities at the end of solidification by the following formula:

$$C_{seg} = \frac{C_{max} - C_{min}}{C_{av}} \tag{3.2}$$

where C_{seg} is the elemental micro-segregation level, C_{max} is the maximum elemental solid solubility at the end of solidification, C_{min} is the minimum solid solubility in the initial stage of solidification, and C_{ave} is the average elemental content (solid solubility) (see Examples 3.6, 3.7 and 3.8).

In order to satisfy the microstructural requirements in good castings, the micro-segregation level in the alloy should be kept at a minimum. Recently (2023), Lu and co-researchers have studied the effects of cooling rates on the solidification behavior, microstructural evolution and mechanical properties of Al–5.5Zn–2,2 Mg–1,2Cu (wt%) alloy. They have reported that high elemental solid solubility and low elemental micro-segregation were achieved at extremely low cooling rates (Lu et al., 2023).

3.3   Cast Structure of Solidified Alloys

Most alloys freeze (solidify) over a temperature range, called the *freezing range* of the alloy. The freezing range of an alloy can be expressed as:

$$\Delta T = T_L - T_S \tag{3.3}$$

where ΔT is the freezing range, T_L is the liquidus temperature, and T_S is the solidus temperature. Figure 3.3 shows the phase diagram of a Ni-Cu alloy that illustrates solidification of the solid-solution binary alloy.

The nickel-copper phase diagram (Fig. 3.3) shows cooling of a 50%Ni-50%Cu alloy (X) from the liquid phase. The *liquidus* curve gives the temperature of the liquid phase at which, for a given composition, the first solid forms. The *solidus* gives the temperature of the two-phase mixture at which the last liquid solidifies. Below the solidus line, the alloy is a single phase: α-solid solution. It has been learnt in Sect. 3.1 that the solidification mechanism involves a dendritic (crystal) growth [see Fig. 3.1(a–c)]. When the 50Ni-50Cu is cooled below the liquidus line,

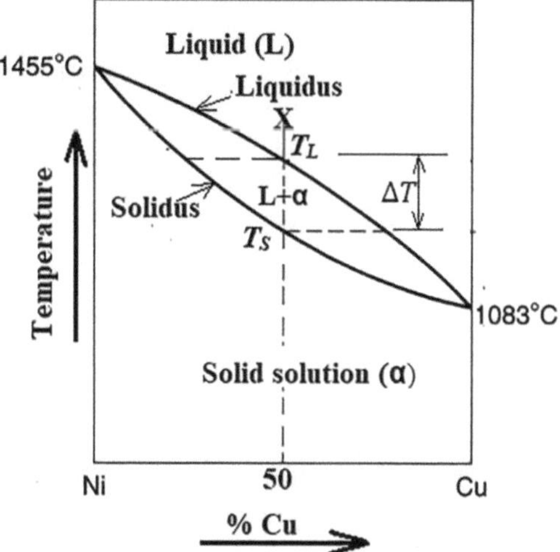

Fig. 3.3 Nickel-copper phase diagram ($\Delta T = T_L - T_S$ = freezing range) (T_L = liquidus temperature, T_S = solidus temperature)

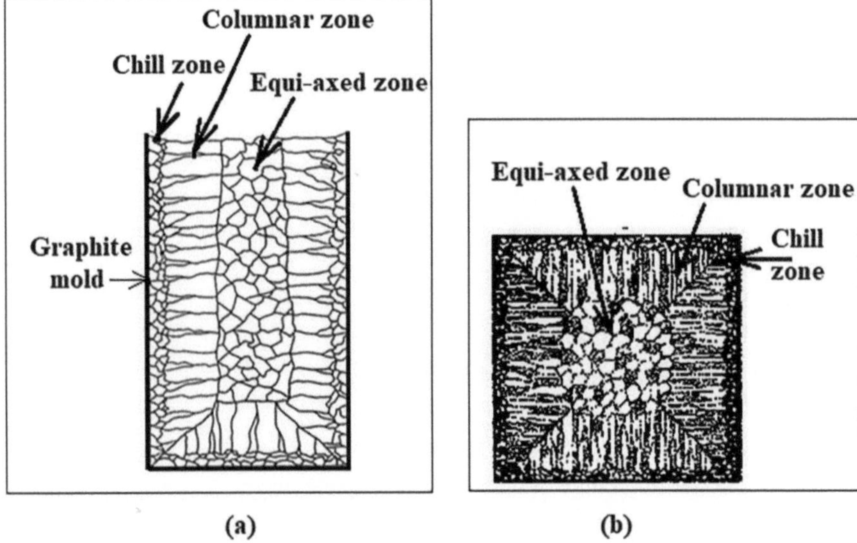

Fig. 3.4 The three-zoned structure of a solid-solution alloy cast in a square graphite mold: (**a**) front view, (**b**) top view

dendrites become richer in Ni, i.e., the first solid (core) is Ni-rich. The alloy is in a two-phase (L + α) region during the freezing range ($\Delta T = T_L - T_S$). On cooling just below the solidus line, the outer region gets enriched in Cu. The solidified structure results in a columnar (mushy) structure with segregation that is non-homogeneous. Thus, non-equilibrium cooling of a solid-solution alloy results in a *cored* (non-homogeneous) structure. The greater the freezing range, the greater will be the non-homogeneity or *coring effect* in the structure (see Example 3.9). The *cored* structure is metallurgically undesirable; it requires homogenization annealing (heat) treatment to obtain a uniform composition throughout the material.

When a solid-solution alloy is cast in a square graphite mold, the material adjacent to the mold-wall cools rapidly to form a fine-grained chill structure (see Fig. 3.4). As the alloy cools away from the mold wall, crystallization proceeds resulting in a (mushy) columnar structure. Finally, the structure is equi-axed in the central zone. Since columnar structure is usually undesirable, nucleating agents (special chemicals) are added to the poured alloy to transform columnar grains to equi-axed grains throughout the material.

3.4   Cooling Curve: Latent Heat of Solidification

The cooling behavior of a pure metal can be illustrated by the cooling curve, as shown in Fig. 3.5. The cooling curve (Fig. 3.5) indicates that a pure metal solidifies at a constant temperature that is equal to its freezing temperature, which is the same

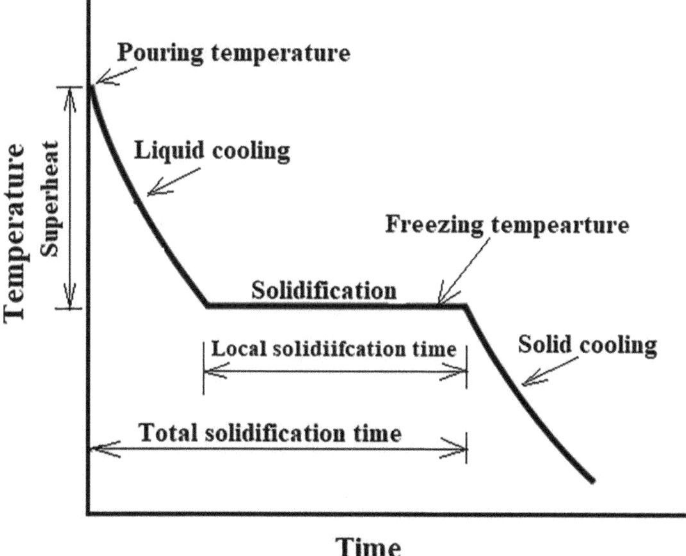

Fig. 3.5 Cooling curve showing solidification behavior of a pure metal

as its melting temperature. The freezing temperature remains constant during the phase change (liquid to solid); the amount of heat energy released during the phase change is equal to the latent heat of solidification, which is the same as the latent heat of fusion.

It has been emphasized in the preceding chapter that the metal (for casting) is heated to a pouring temperature that is reasonably higher than the melting temperature. The difference between the pouring temperature and the melting temperature of a metal is called *superheat* (see Fig 3.5). The *total solidification time, t_s,* is the sum of pouring time and local solidification time. It means that the total solidification time, t_s, is the time required for the casting from pouring to complete solidification. The time t_s depends on the size and shape of the casting.

3.5   Heat Transfer in Solidification: Fick's Second Law

This section describes heat transfer and solidification in an insulating (sand or ceramic) mold. Consider pouring of a pure metal into a sand mold at a pouring temperature, T_p. At some time, t, after pouring, the temperature along a line perpendicular to the mold-metal interface can be graphically represented, as shown in Fig. 3.6.

It is evident in Fig. 3.6 that firstly, the liquid metal (with high thermal conductivity) cools instantly from its pouring temperature, T_p, to its melting temperature, T_m upon pouring. The heat transfer in the mold can be expressed by Fick's second law, as follows:

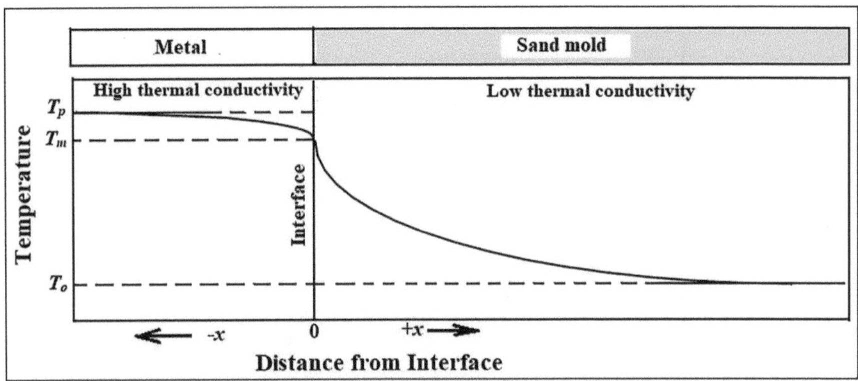

Fig. 3.6 The variation of temperature with the distance in a sand mold

$$\frac{\partial T}{\partial t} = \alpha_M \bullet \frac{\partial^2 T}{\partial^2 x} \tag{3.4}$$

where $\dfrac{\partial T}{\partial t}$ is the rate of heat transfer, x is the distance from metal-mold interface (m), and α_M is the thermal diffusivity of mold (m²s⁻¹), which can be expressed by (Zanganeh et al., 2013):

$$\alpha_M = \frac{K_M}{\rho_M \bullet C_M} \tag{3.5}$$

where K_M is the thermal conductivity of the mold (W·m⁻¹·°C⁻¹), ρ_M is the density of the mold (kg •m⁻³), and C_M is the specific heat of the mold material at constant pressure (J •kg⁻¹ °C⁻¹). It has been experimentally shown that the thermal diffusivity of foundry mold made of used green sand (silica quartz +5 wt% bentonite +3 wt% water) lies in the range of 2.0×10^{-7}–2.0×10^{-6} m²s⁻¹ for the temperature in the range of 20–500 °C (Krajewski et al., 2016). Table 3.1 lists important thermal properties of some commonly used mold materials in a foundry. The significance of Eq. 3.5 is illustrated in Examples 3.10 and 3.11.

Engineering analysis of heat transfer in solidification enables us to derive a workable solution to the differential equation: Eq. 3.4. It can be shown by use of Calculus that the total solidification time of a casting can be expressed as:

$$t_s = \frac{\pi \bullet \alpha_M \bullet \rho_{\text{Metal}}^2}{4 K_M^2 \bullet \left(T_m - T_o\right)^2} \bullet \left[\Delta H_f + C_{\text{Metal}}\left(T_p - T_m\right)\right]^2 \bullet \left(\frac{V_{\text{Metal}}}{A}\right)^2 \tag{3.6}$$

where t_s is the total solidification time (s), ΔH_f is the heat of fusion (J•kg⁻¹), $(T_p - T_m)$ is the superheat (°C), V_{Metal} is the volume of the casting (m³), and A is the surface area of the casting through which heat is extracted (m²).

Table 3.1 Important physical/thermal properties of some foundry materials

Material	Density, ρ (kg $\bullet$m^{-3})	Specific heat, C (J $\bullet$kg^{-1}°C^{-1})	Thermal conductivity, K (W·m^{-1} · °C^{-1})	Heat of fusion, ΔH_f (J $\bullet$kg^{-1})
Silica	1600 (mold)	1117	1.5	–
	2320 (bulk)			
Fe (liq)	7015	795	–	272090
Fe (s) 20 °C	7870	456	78.2	272090
Al (liq)	2375	1080	94.0	395995
Al (s) 20 °C	2700	918	238	395995
Cu (liq)	8000	495	166	211812
Cu (s) 20 °C	8960	386	397	211812

Equation 3.6, in its simplest form, can be expressed as:

$$t_s = C_m \bullet \left(\frac{V_{\text{Metal}}}{A} \right)^2 \tag{3.7}$$

where C_m is the mold constant given by:

$$C_m = \frac{\pi \bullet \alpha_M \bullet \rho_{\text{Metal}}^2}{4 K_M^2 \bullet \left(T_m - T_o\right)^2} \bullet \left[\Delta H_f + C_{\text{Metal}} \left(T_p - T_m\right) \right]^2 \tag{3.8}$$

The significance of Eq. 3.8 is illustrated in Examples 3.12, 3.13 and 3.14.

Equation 3.7 is usually written in its generalized form that involves a constant n, which is discussed in the following section.

3.6   Chvorinov's Rule

It has been mentioned in the preceding section that the solution of Fick's law enables us to obtain a workable expression for determining the solidification time for a casting operation. The total solidification time is related to the casting geometry by *Chvorinov's rule,* which is expressed by (Campbell, 2011):

$$t_s = C_m \bullet \left(\frac{V}{A} \right)^n \tag{3.9}$$

where t_s is the total solidification time (min), C_m is the mold constant (min/cm^2), V is the volume of the casting (cm^3), A is the surface area through which heat is extracted (cm^2), and n is a constant ($n = 1.5$–2.0). The terms in Eq. 3.9 can be rearranged for determining the mold constant, as follows:

$$C_m = t_s \bullet \left(\frac{A}{V}\right)^n \tag{3.10}$$

In most cases, the value of the constant $n = 2$. Hence, Eq. 3.10 may be rewritten as:

$$t_s = C_m \bullet \left(\frac{V}{A}\right)^2 \tag{3.11}$$

Chvorinov's rule implies that a casting with a higher volume-to-surface area ratio cools and solidifies more slowly than one with a lower ratio.

It must be noted that Eqs. 3.9, 3.10 and 3.11 are applicable to only those casting that involve one-dimensional heat transfer (e.g., plates, sheets). This limitation has been overcome by Jelínek and Elbel (2010), who have generalized *Chvorinov's rule* to include heat transfer on concave mold surfaces curved inward (e.g., a sphere, a cylinder, etc.). The generalized expression is given by (Jelínek & Elbel, 2010):

$$t_s = \varepsilon \bullet C_m \bullet \left(\frac{V}{A}\right)^2 \tag{3.12}$$

where ε is the shape coefficient. For a plate form casting $\varepsilon = 1$, i.e., Eq. 3.12 would simplify to Eq. 3.11; for other forms (cylinder, sphere, etc.), $\varepsilon < 1$. The significances of Eqs. 3.9, 3.10, 3.11 and 3.12 are illustrated in Examples 3.15, 3.16, 3.17, 3.18, 3.19 and 3.20.

3.7 Worked Numerical Examples in Solidification Design

Example 3.1 Drawing a Graphical Plot for Cooling Rate-Grain Size Data
The effect of cooling rate on the austenite grain size of an alloy-steel (composition in wt%: 58.7Fe-30Mn-10Al-1.1C, approx.), during its solidification, was observed. The kinetic data, so obtained, is presented in Table 3.2.

(a) Plot a graph for the data in Table 3.2. Take $\dot{\theta}$ as abscissa and d as ordinate.
(b) What solidification behavior is reflected by the graph drawn by you in (a)?

Solution
(a) The graphical plot, for the data in Table 3.2, is shown in Fig. 3.7.
(b) The grain size decreases with increase in the cooling rate.

Table 3.2 The data for Example 3.1

Cooling rate, $\dot{\theta}$ (°C/s)	6	7	8	9	10
Av. grain size, d (μm)	109	104	99.5	94.5	89.5

Fig. 3.7 Graphical plot for
Example 3.1

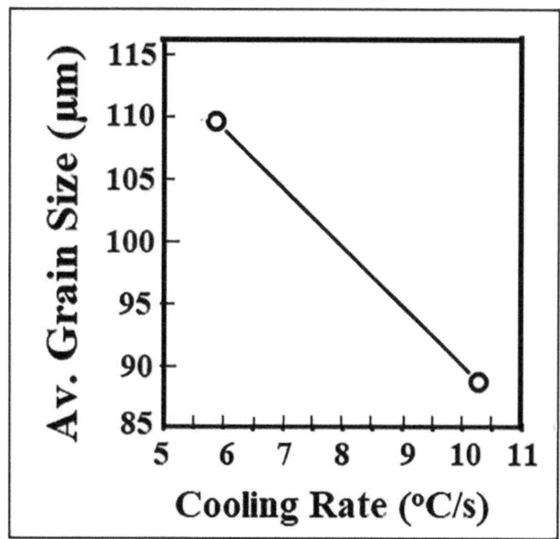

Example 3.2 Determining the Grain Sizes for Given Cooling Rates

By reference to the graph drawn in Example 3.1, determine the grain sizes for the
following cooling rates: (a) 6.5 °C/s, and (b) 9.5 °C/s. Clearly mark your answers on
the graph.

Solution

The corresponding grain sizes for the required cooling rates are shown in Fig. 3.8.

(a) For the cooling rate of 6.5 °C/s, the grain size = 106.5 µm,
(b) For the cooling rate of 9.5 °C/s, the grain size = 92 µm

Example 3.3 Determining the Slope of the Curve/Line for Cooling Rate-Grain Size Plot

By using the data in Example 3.2, determine the slope of the curve/line in Fig. 3.8.

Solution

By reference to Fig. 3.8,

$$\Delta \dot{\theta} = 9.5\text{–}6.5 = 3 \text{ °C/s, and } \Delta d = 92\text{–}106.5 = -14.5 \text{ µm.}$$

$$\text{Slope} = \frac{\Delta d}{\Delta \dot{\theta}} = \frac{-14.5}{3} = -4.833$$

Example 3.4 Development of an Equation for Cooling Rate-Grain Size Relationship

Develop the solidification-design equation for the graph drawn by you in
Example 3.1.

Fig. 3.8 Graphical plot for Example 3.2

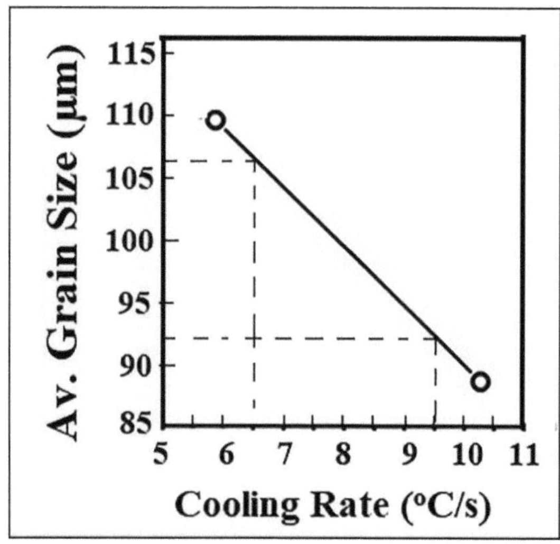

Solution

Slope = −4.833, d-intercept = 114.

The required equation, in the slope-intercept form, can now be developed as follows:

$$d = -4.833 \cdot \dot{\theta} + 114$$

where d is the average grain size (μm) and $\dot{\theta}$ is cooling rate (°C/s).

Example 3.5 Determining the Cooling Rate by Using the Solidification-Design Equation

It is required to achieve an austenite grain size of 90 μm in the cast (58.7Fe-30Mn-10Al-1.1C, approx.) alloy steel. Determine the cooling rate by using the equation developed in Example 3.4.

Solution

By substituting d = 90 in the equation: d = −4.833 $\bullet\dot{\theta}$ + 114.

$$90 = -4.833 \cdot \dot{\theta} + 114$$
$$-4.833 \cdot \dot{\theta} = -24$$

$$\text{Cooling rate} = \dot{\theta} = 4.96° \text{C} / \text{s} \cong 5° \text{C} / \text{s}$$

Example 3.6 Determining the Average Elemental (Zn) Content in a Solidified Alloy

An Al-Zn-Mg-Cu alloy was solidified at different cooling rates and the maximum and minimum Zn contents (solid solubility) were recorded, as shown in Table 3.3.

Table 3.3 The Zn contents in the Al-Zn-Mg-Cu alloy at different cooling rates

Cooling rate (K/s)	C_{min} (wt%)	C_{max} (wt%)	C_{av} (wt%)
0.3	4.36	6.31	
3.4	3.52	7.01	
10.4	3.24	7.74	

Table 3.4 The average Zn contents in the Al-Zn-Mg-Cu alloy at different cooling rates

Cooling rate (K/s)	C_{min} (wt%)	C_{max} (wt%)	C_{av} (wt%)
0.3	4.36	6.31	5.33
3.4	3.52	7.01	5.25
10.4	3.24	7.74	5.49

Determine the average Zn content for each cooling rate, and complete the data in Table 3.3 to form a new table.

Solution

For the cooling rate of 0.3 K/s,

$$C_{av} = \frac{C_{min} + C_{max}}{2} = \frac{4.36 + 6.31}{2} = 5.33 \ \text{wt}\%$$

For the cooling rate of 3.4 K/s,

$$C_{av} = \frac{3.52 + 7.01}{2} = 5.25 \ \text{wt}\%$$

For the cooling rate of 10.4 K/s,

$$C_{av} = \frac{3.24 + 7.74}{2} = 5.49 \ \text{wt}\%$$

The completed data is shown in Table 3.4.

Example 3.7 Determining the Micro-segregation Level at Different Solidification Rates

By using the data in Example 3.6, determine the micro-segregation level in the alloy for each cooling rate shown in Table 3.4. Show your answers as a new Table 3.5.

Solution

By using Eq. 3.2 for the cooling rate data in Table 3.4,

For the cooling rate of 0.3 K/s,

$$C_{seg} = \frac{C_{max} - C_{min}}{C_{av}} = \frac{6.31 - 4.36}{5.33} = 0.366$$

For the cooling rate of 3.4 K/s,

$$C_{seg} = \frac{7.01 - 3.52}{5.25} = 0.665$$

For the cooling rate of 10.4 K/s,

$$C_{seg} = \frac{7.74 - 3.24}{5.49} = 0.820$$

The micro-segregation data, so obtained, is shown in Table 3.5.

Example 3.8 Drawing a Column Chart for Cooling Rate-Micro-segregation Data

Plot a column chart by using the data in Table 3.5 (take the cooling rates as abscissa).

What do you conclude from the graphical plot? Is the solidification behavior usual or exceptional?

Solution

The column chart, for the data in Table 3.5, is shown in Fig. 3.9.

It can be concluded that the lowest elemental micro-segregation were achieved at the lowest (extremely low) cooling rate. The solidification behavior is exceptional.

Table 3.5 The micro-segregation data for various cooling rates for the Al-Zn-Mg-Cu alloy

Cooling rate (K/s)	C_{seg}
0.3	0.366
3.4	0.665
10.4	0.820

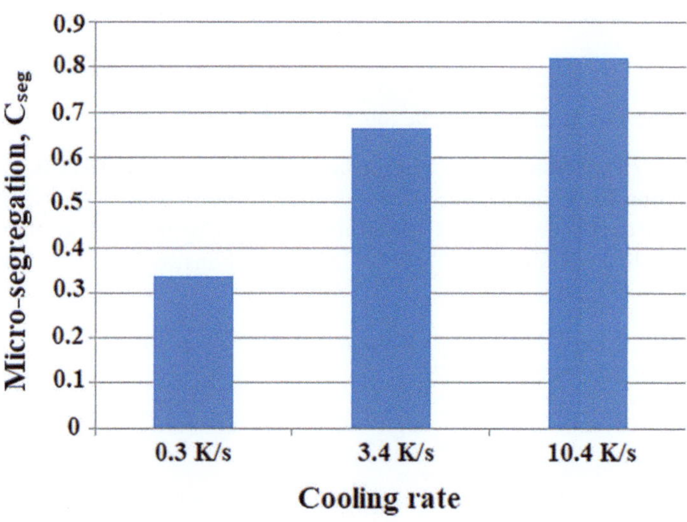

Fig. 3.9 The column chart for Example 3.8

Example 3.9 Identifying Severity of Coring Effect by Determining the Freezing Range

Two nickel-copper alloys (Ni-10Cu and Ni-50Cu) are poured at a temperature of 1500 °C in two separate casting experiments. Which alloy would likely develop greater coring effect in the cast structure after solidification? Hint: refer to Fig. 3.3.

Solution

By reference to Fig 3.3, a (new) labeled phase diagram is shown in Fig. 3.10.
With reference to Fig. 3.10 and by using Eq. 3.3, we get:

$$\text{For Ni-10Cu alloy,} \quad \Delta T = T_L - T_S = 1400 - 1250 = 150°C$$

$$\text{For Ni-50Cu alloy,} \quad \Delta T = 1050 - 800 = 250°C$$

Since the freezing range (ΔT) is greater in case of Ni-50Cu alloy, the alloy is likely to develop greater coring effect after solidification.

Example 3.10 Determining the Thermal Diffusivity of Silica (Molding) Sand

A metallic component is required to be sand (silica) cast. Determine the thermal diffusivity of the molding sand. Hint: use the data in Table 3.1.

Solution

By reference to Table 3.1,
$\rho_M = 1600$ kg $\bullet$m^{-3}, and $C_M = 1117$ J $\bullet$kg^{-1}°C^{-1}, $K_M = 1.5$ W·m^{-1}·°C^{-1}, $\alpha_M = ?$
By using Eq. 3.5,

$$\alpha_M = \frac{K_M}{\rho_M \cdot C_M} = \frac{1.5}{1600 \times 1117} = 839.3 \times 10^{-9} = 8.39 \times 10^{-7} \ \text{m}^2\text{s}^{-1}$$

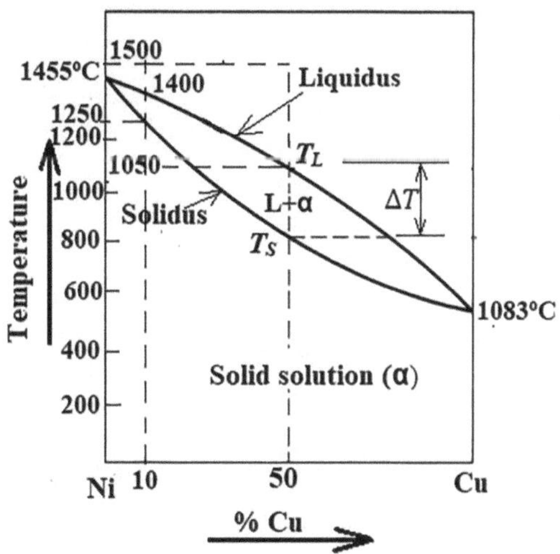

Fig. 3.10 Phase diagram for Example 3.9

The thermal diffusivity of silica mold = 8.39×10^{-7} m²s⁻¹

Example 3.11 Determining the Mold Constant for Steel Casting with Superheat of 50 °C

Molten mild steel has been poured, with a superheat of 50 °C, into a (silica) sand mold. Determine the mold constant. Take the melting point of the steel = 1370 °C.

Solution

By reference to *Example 3.10,* and the data in Table 3.1,

$\alpha_M = 8.39 \times 10^{-7}$ m²s⁻¹, $K_M = 1.5$ W·m⁻¹· °C⁻¹, $\rho_{\text{Metal}} = 7015$ kg•m⁻³, $C_{\text{Metal}} = 795$ J•kg⁻¹°C⁻¹

$T_0 = 20$ °C, $T_p - T_m = 50$ °C, $T_m = 1370$ °C, $\Delta H_f = 272090$ J•kg⁻¹, $C_m = ?$

By using Eq. 3.8,

$$C_m = \frac{\pi \cdot \alpha_M \cdot \rho_{\text{Metal}}^2}{4 K_M^2 \cdot \left(T_m - T_o\right)^2} \cdot \left[\Delta H_f + C_{\text{Metal}}\left(T_p - T_m\right)\right]^2$$

$$C_m = \frac{\pi \times 8.39 \times 10^{-7} \times 7015^2}{4 \times 1.5^2 \times \left(1370 - 20\right)^2} \times \left(272090 + 795 \times 50\right)^2$$

$$= \frac{129.71}{4 \times 2.25 \times 1822500} \times 311840^2$$

$$C_m = 7.91 \times 10^{-6} \times 9.72 \times 10^{10} = 768852 \text{ s} \cdot \text{m}^{-2}$$

The mold constant = $C_m = 768852$ s •m⁻² = 1.28 min •cm⁻²

Example 3.12 Determining and Comparing Cm for Steel Casting with Superheat of 200 °C

Repeat Example 3.11 when steel is poured with superheat of 200 °C. Determine the percent difference, and comment on the effect of superheat on the mold constant.

Solution

$\alpha_M = 8.39 \times 10^{-7}$ m²s⁻¹, $K_M = 1.5$ W·m⁻¹· °C⁻¹, $\rho_{\text{Metal}} = 7015$ kg•m⁻³, $C_{\text{Metal}} = 795$ J•kg⁻¹°C⁻¹

$T_0 = 20$ °C, $T_m = 1370$ °C, $T_p - T_m = 200$ °C, $\Delta H_f = 272090$ J•kg⁻¹, $C_m = ?$

By using Eq. 3.8,

$$C_m = \frac{\pi \times 8.39 \times 10^{-7} \times 7015^2}{4 \times 1.5^2 \times \left(1370 - 20\right)^2} \times \left(272090 + 795 \times 200\right)^2$$

$$= \frac{129.71}{4 \times 2.25 \times 1822500} \times 431090^2$$

$$C_m = 7.91 \times 10^{-6} \times 1.858 \times 10^{11}$$
$$= 1469678 \text{ s} \cdot \text{m}^{-2} = 2.45 \text{ min} \cdot \text{cm}^{-2}$$

For 50 °C superheat, C_m = 1.28 min•cm^{-2}, for 200 °C superheat, C_m = 2.45 min•cm^{-2}

$$\%\text{Difference} = \frac{2.45 - 1.28}{1.28} \times 100 = 91.4$$

When the superheat is increased from 50 to 200 °C, the mold constant is increased by 91.4%; it means that the casting will take 91.4% longer total time to solidify.

Example 3.13 Determining and Comparing the Mold Constant for Aluminum Casting

Repeat Example 3.11 if the metal cast is aluminum instead of steel. Determine the percent difference, and comment on the effect of metal cast on the mold constant.

Solution

By reference to the data for aluminum in Table 3.1,

α_M = 8.39 × 10^{-7} m^{2}s^{-1}, K_M= 1.5 W·m^{-1}· °C^{-1}, ρ_{Metal} = 2,375 kg•m^{-3}, C_{Metal} = 1080 J•kg^{-1}°C^{-1}

T_o = 20 °C, $T_p - T_m$ = 50 °C, T_m = 660 °C, ΔH_f= 395995 J•kg^{-1}, C_m =?

By using Eq. 3.8,

$$C_m = \frac{\pi \cdot \alpha_M \cdot \rho_{\text{Metal}}^2}{4 K_M^2 \cdot (T_m - T_o)^2} \cdot \left[\Delta H_f + C_{\text{Metal}} (T_p - T_m) \right]^2$$

$$C_m = \frac{\pi \times 8.39 \times 10^{-7} \times 2375^2}{4 \times 1.5^2 \times (660 - 20)^2} \times (395995 + 1080 \times 50)^2$$

$$= \frac{14.867}{4 \times 2.25 \times 409600} \times 449995^2$$

$$C_m = 4.033 \times 10^{-6} \times 2.025 \times 10^{11}$$
$$= 816682.50 \text{ s} \cdot \text{m}^{-2} = 1.36 \text{ min} \cdot \text{cm}^{-2}$$

For steel casting, C_m = 1.28 min•cm^{-2}; for aluminum casting, C_m = 1.36 min•cm^{-2}

$$\% \text{ Difference} = \frac{1.36 - 1.28}{1.28} \times 100 = 6.25$$

When the casting metal changes from mild steel to aluminum, the mold constant is increased by 6.25%; it means that the casting will take 6.25% longer total time to solidify.

Example 3.14 Determining the Total Solidification Time for a Casting
A rectangular aluminum product (14 cm × 8 cm × 3 cm) is to be cast by sand casting process. Determine the total solidification time for the casting, if the mold constant of Chvorinov's rule is 3.5 min/cm². Take n = 2.

Solution

$$\text{Surface area} = A = 2\left[(14\times8)+(14\times3)+(8\times3)\right]$$
$$= 2\times(112+42+24) = 356 \ \text{cm}^2$$

$$\text{Volume} = V = 14\times8\times3 = 336 \ \text{cm}^3$$

By using Eq. 3.11,

$$t_s = C_m\left(\frac{V}{A}\right)^2 = 3.5\times\left(\frac{336}{356}\right)^2$$
$$= 3.5\times(0.9438)^2 = 3.12 \ \text{min}$$

The total solidification time = 3.12 min.

Example 3.15 Determining the Mold Constant of Chvorinov's rule
A cast iron plate has dimensions: 17 cm × 13 cm × 4 cm. The plate was produced by sand casting. The total solidification time was 5 min. Determine the mold constant of Chvorinov's rule. Take n = 1.8.

Solution
t_s = 3 min, n = 1.8, Volume = V = 17 cm × 13 cm × 4 cm = 884 cm³

$$\text{Surface area} = A = 2\left[(17\times13)+(13\times4)+(17\times4)\right]$$
$$= 2(221+52+68) = 682 \ \text{cm}^2$$

By using Eq. 3.9,

$$C_m = t_s\cdot\left(\frac{A}{V}\right)^n = 3\times\left(\frac{682}{884}\right)^{1.8}$$
$$= 3\times(0.771)^{1.8} = 1.88 \ \text{min/cm}^2$$

The mold constant = 1.88 min/cm².

Example 3.16 Determining the Solidification Time for a Casting When Cm Is Unknown

A metal plate casting having dimensions (20 cm × 16 cm × 4 cm) solidifies in 7 min. Calculate the total solidification time for another casting with dimensions: (18 cm × 11 cm × 3 cm); the same metal is poured under the same conditions.

Solution

For casting with dimensions (20 cm × 16 cm × 4 cm)

$t_s = 7$ min, $V = 20$ cm × 16 cm × 4 cm = 1280 cm^3

$$\text{Surface area} = A = 2\left\{(20\times16)+(20\times4)+(16\times4)\right\}$$
$$= 2\times(320+80+64) = 928 \text{ cm}^2$$

By using Eq. 3.9,

$$C_m = t_s \cdot \left(\frac{A}{V}\right)^n = 7\times\left(\frac{928}{1280}\right)^2$$
$$= 7\times(0.725)^2 = 3.68 \text{ min/cm}^2$$

For casting with dimensions (18 cm × 11 cm × 3 cm)

$C_m = 3.68$ min/cm^2, $V = 18$ cm × 11 cm × 3 cm = 594 cm^3

$$A = 2\left\{(18\times11)+(18\times3)+(11\times3)\right\}$$
$$= 2(198+54+33) = 570 \text{ cm}^2$$

By using Eq. 3.11,

$$t_s = C_m\left(\frac{V}{A}\right)^2 = 3.68\times\left(\frac{594}{570}\right)^2$$
$$= 3.68\times(1.042)^2 = 3.996 \text{ min}$$

The total solidification time for casting with dimensions (18 cm × 11 cm × 3 cm) = 4 min.

Example 3.17 Determining the Total Solidification Time for a Spherical Metallic Casting

A metallic solid sphere with 8 cm diameter is to be cast. The shape coefficient is 0.9 and the mold constant is 3.6 min/cm^2. Determine the total solidification time for the casting operation.

Solution

$C_m = 3.6$ min/cm^2, $\varepsilon = 0.9$, radius = r = 4 cm

$$\text{Volume of sphere} = V = \frac{4}{3} \cdot \pi \cdot r^3$$

$$= 1.33 \times \pi \times 4^3 = 267.4 \ \text{cm}^3$$

$$\text{Surface area of sphere} = A = 4 \cdot \pi \cdot r^2$$

$$= 4 \times \pi (4)^2 = 201.062 \ \text{cm}^2$$

By using Eq. 3.12,

$$t_s = \varepsilon \cdot C_m \left(\frac{V}{A}\right)^2 = 0.9 \times 3.6 \times \left(\frac{267.4}{201.062}\right)^2$$

$$= 0.9 \times 3.6 \times (1.3299)^2 = 5.757 \ \text{min} \cong 6 \ \text{min}$$

The total solidification time $\cong$ 6 min.

Example 3.18 Determining the Volume and Surface Area for a Cast Metallic Coin

A metallic coin was cast. The diameter of the coin is 15 cm and its thickness is 3 mm. Determine the volume and the surface area through which heat was extracted.

Solution

D = 15 cm, h = 3 mm = 0.3 cm, V =?, A =?

$$\text{Volume of the cylindrical} \ (\text{disc}) \ \text{coin} = V = \frac{\pi}{4} D^2 \cdot h = \frac{\pi}{4} 15^2 \times 0.3$$

$$= 0.785 \times 225 \times 0.3 = 53 \ \text{cm}^3$$

$$\text{Surface area} = A = \pi \cdot D \cdot h + 2 \left(\frac{\pi}{4} D^2\right)$$

$$= (\pi \times 15 \times 0.3) + 2 \times (0.785 \times 15^2) = 353.25 \ \text{cm}^2$$

Example 3.19 Determining the Mold Constant for a Cast Metallic Coin

By using the data in Example 3.17, determine the mold constant for the casting operation if it took 2 min to totally solidify the casting. Take shape coefficient for solidification as 0.95.

Solution

V = 53 cm³, A = 353.25 cm², t_s = 2 min., ε = 0.95, C_m =?

By using the modified form of Eq. 3.10,

$$c_m = \frac{t_s}{\varepsilon} \cdot \left(\frac{A}{V}\right)^2 = \left(\frac{2}{0.95}\right) \times \left(\frac{353.25}{53}\right)^2$$

$$= 2.105 \times 44.4235 = 46.53 \ \text{min/ cm}^2$$

The mold constant = 46.53 min/cm².

Example 3.20 Comparison of Solidification Times for Various Solid Shapes
Three metallic solid pieces with the same volume are required to be cast. One piece
is a cube, the second is a sphere, and the third is a cylinder with its height equal to
its diameter. The shapes coefficients for the cubic, spherical, and cylindrical pieces
are 1.00, 0.90, and 0.95, respectively. Which piece will solidify the slowest and
which one the fastest? Take n = 2.

Solution
Since the volume is same for all shapes, it can be taken as unity, i.e., V = 1

Cube

$$V = a^3, a^3 = 1, a = 1$$

$$A = 6a^2 = 6 \times 1 = 6$$

Sphere

$$V = \frac{4}{3} \cdot \pi \cdot r^3, \qquad \frac{4}{3} \cdot \pi \cdot r^3 = 1, \qquad r = \left(\frac{3}{4\pi}\right)^{1/3}$$

$$A = 4 \cdot \pi \cdot r^2 = 4 \times \pi \times \left(\frac{3}{4\pi}\right)^{2/3} = 4.84$$

Cylinder

$$h = 2r, V = \pi \cdot r^2 \cdot h = 2 \cdot \pi \cdot r^3, 2 \cdot \pi \cdot r^3 = 1, r = \left(\frac{1}{2 \cdot \pi}\right)^{1/3}$$

$$A = \left(2 \cdot \pi \cdot r^2\right) + \left(2 \cdot \pi \cdot r \cdot h\right) = \left(2 \cdot \pi \cdot r^2\right) + \left(4 \cdot \pi \cdot r^2\right)$$

$$= 6 \cdot \pi \cdot r^2 = 6 \times \pi \times \left(\frac{1}{2 \cdot \pi}\right)^{2/3} = 5.54$$

By using Eq. 3.12,

$$t_{cube} = \varepsilon \cdot C_m \left(\frac{V}{A}\right)^2 = 1 \times C_m \left(\frac{1}{6}\right)^2 = 0.0277\ C_m$$

$$t_{sphere} = C_m \left(\frac{V}{A}\right)^2 = 0.9 \times C_m \left(\frac{1}{4.84}\right)^2 = 0.0384\ C_m$$

$$t_{cylinder} = 0.95 \times C_m \left(\frac{1}{5.54}\right)^2 = 0.0309\ C_m$$

The spherical casting will solidify the slowest, whereas the cubic casting will solidify the fastest.

Questions and Problems

3.1. Encircle the most appropriate answers for each of the following statements.

(1) Which stage of solidification involves formation of grained microstructure?

(a) Nucleation, (b) growth of nuclei, (c) progress of crystallization, (d) completion of solidification

(2) Which stage of solidification involves formation of multiple nuclei in the melt?

(a) Nucleation, (b) growth of nuclei, (c) progress of crystallization, (d) completion of solidification

(3) Which mechanism of solidification results in a fine-grained microstructure?

(a) The rate of growth exceeds the rate of nucleation (G > N), (b) N > G, (c) N = G

(4) Which law/rule governs the relationship between grain size and yield strength?

(a) Chvorinov's rule, (b) Fick's law, (c) Hall–Petch law, (d) Bernoulli's law

(5) In which zone is formed a fine-grained microstructure in an alloy cast in a square graphite mold?

(a) Chill zone, (b) columnar zone, (c) equi-axed zone

(6) Which cooling rate generally results in a fine-grained microstructure and low elemental micro-segregation?

(a) Low cooling rate, (b) high cooling rate, (c) very low cooling rate

(7) Which cooling rate results in a low elemental micro-segregation in Al–5.5Zn–2.2Mg–1.2Cu (wt%) alloy?

(a) Low cooling rate, (b) high cooling rate, (c) very low cooling rate

(8) The difference of pouring temperature and the freezing temperature is called ______

(a) Freezing range, (b) specific temperature, (c) super-temperature, (d) super-heat
(9) Which type of casting will cool and solidify more slowly?
(a) Higher (V/A) ratio, (b) lower (V/A) ratio, (c) very low (V/A) ratio
(10) Which of the following shapes will solidify the fastest during metal casting?
(a) Cubic, (b) spherical, (c) cylindrical

3.2. Draw a labeled sketch showing the steps in crystallization during solidification of metal.

3.3. Why does a high cooling rate, during solidification of metal, result in a fine-grained microstructure?

3.4. (a) Explain the *coring* effect in a solid-solution alloy when solidified under non-equilibrium condition.

(b) How would you overcome the problem of "cored" structure?

3.5. Draw a labeled cooling curve showing solidification behavior of a pure metal.

P-3.6. An Al-Zn-Mg-Cu alloy was solidified at different cooling rates and the maximum and the minimum Mg contents (solid solubility) were recorded, as shown in Table 3.6(a).

(a) Determine the average Mg content for each cooling rate, and complete the data in Table 3.6(a) to form a new table - Table 3.6(b).

(b) Determine the micro-segregation level of Mg in the alloy for each cooling rate shown in Table 3.6(b). Show your answers as a new table - Table 3.6(c).

P-3.7. Plot a column chart by using the data in Table 3.6c (take the cooling rates as abscissa). What do you conclude from the graphical plot? Is the solidification behavior usual or exceptional?

P-3.8. Molten mild steel has been poured, with a superheat of 100 °C, into a (silica) sand mold. Determine the mold constant. Hint: refer to the data in Table 3.1.

P-3.9. Repeat Problem P-3.8 when steel is poured with superheat of 300 °C. Determine the percent difference, and comment on the effect of super-heat on the mold constant.

P-3.10. A rectangular cast iron piece (16 cm × 10 cm × 4 cm) is to be manufactured by sand casting process. Determine the total solidification time for the casting, if the mold constant of Chvorinov's rule is 3.2 min/cm². Take n = 2.

P-3.11. A rectangular metallic casting with dimensions (18 cm × 13 cm × 6 cm) solidifies in 6 min. Calculate the total solidification

Table 3.6 (a) Mg contents in the Al-Zn-Mg-Cu alloy

Cooling rate (K/s)	C(min) (wt%)	C(max) (wt%)	C(av)
0.3	1.87	2.81	
3.4	1.37	3.36	
10.4	1.24	3.72	

time for another casting with dimensions: (14 cm × 10 cm × 4 cm); the same metal is poured under the same conditions.

P-3.12. A metallic solid sphere with 10 cm diameter is to be cast. The shape coefficient is 0.9 and the mold constant is 3.5 min/cm². Determine the total solidification time for the casting operation.

P-3.13. A metallic coin is to be cast. The diameter of the coin is 10 cm and its thickness is 2 mm. Determine the total solidification time, if the mold constant is 10 min/cm².

References

Andresen, W. (2004). *Die cast engineering: A hydraulic, thermal, and mechanical process.* CRC Press.

Campbell, J. (2011). *Complete casting handbook: Metal casting processes, techniques and design.* Butterworth-Heinemann/Elsevier Science Publications.

Cui, Z. X., & Qin, Y. C. (2007). *Metallography and heat treatment* (2nd ed.). Beijing.

Hansen, N. (2004). Hall–Petch relation and boundary strengthening. *Scripta Materilia, 51,* 801–806.

He, C., Yu, W., Li, Y., Wang, Z., Wu, D., et al. (2020). Relationship between cooling rate, microstructure evolution, and performance improvement of an Al–Cu alloy prepared using different methods. *Materials Research Express, 7,* 116501.

Huang, S., Li, G., Zhang, Z., et al. (2022). Effect of cooling rate on the grain morphology and element segregation behavior of Fe-Mn-Al-C low-density steel during solidification. *Processes, 10*(6), 1101.

Huda, Z. (2020). *Metallurgy for physicists and engineers.* CRC Press.

Jelínek, P., & Elbel, T. (2010). Chvorinov's rule and determination of coefficient of heat accumulation of molds with non-quartz base sands. *Archives of Foundry Engineering, 10*(4), 77–82.

Krajewski, P. K., Piwowarski, G., & Buraś, J. (2016). Thermal properties of foundry mould made of used green sand. *Archives of Foundry Engineering, 16*(1), 29–32.

Liu, Z.-T., Wang, B.-Y., Wang, C., Zha, M., et al. (2020). Microstructure and mechanical properties of Al-Mg-Si alloy fabricated by a short process based on sub-rapid solidification. *Journal of Materials Science & Technology, 41,* 178–186.

Lu, B., Li, Y., Wang, H., et al. (2023). Effects of cooling rates on the solidification behavior, microstructural evolution and mechanical properties of Al–Zn–Mg–Cu alloys. *Journal of Materials Research and Technology, 22,* 2532–2548.

Tang, H.-P., Wang, Q.-D., Lei, C., et al. (2019). Effect of cooling rate on microstructure and mechanical properties of an Al-5.0Mg-3.0Zn-1.0Cu cast alloy. *Journal of Alloys and Compounds, 801,* 596–608.

Zanganeh, J., Moghtaderi, B., & Ishida, H. (2013). Combustion and flame spread on fuel-soaked porous solids. *Progress in Energy and Combustion Science, 39*(4), 320–339.

Chapter 4
Melting Practice and Furnaces

Nomenclature

V_s	Secondary voltage from transformer
R_L	Load resistance
R_s	Source resistance
X_s	Source reactance
X_L	Load reactance
P_{peak}	Peak power to the arc
P_g	Gross power (of induction furnace)
i	Hot metal ratio
M_{EAF}^{i}	Carbon emission of power per ton of molten steel for the hot metal ratio i in EAF
Q_{EE}^{i}	Electrical energy consumption per ton of molten steel for the hot metal ratio i in EAF
C_{Scoal}	Carbon content of standard coal
m_{additive}	Mass of the element required to be added
m_{ladle}	Mass of the molten steel in the ladle
$\Delta\%X$	Required increase in the $wt.\%$ of the alloying element X
$\%X_{\text{aim}}$	Aim (target) wt.% of element X in the alloy
$\%X_{\text{actual}}$	Actual wt.% of element X in the alloy (from spectrometry result)
$m_{Fe-\text{alloy}}$	Mass of the ferroalloy
Δ wt.% C	Amount of carbon pick up (wt.%) by steel
ΔH_f	Heat of fusion
m_{scrap}	Mass of scrap to be added
$\text{Capacity}_{\text{EAF}}$	Capacity of EAF (tons/heat)
v	Casting speed of CC machine

Z. Huda, *Metal Casting Engineering*, Mechanical Engineering Series,
https://doi.org/10.1007/978-3-031-84620-5_4

Abbreviations

AMF   American Foundry Society
EAF   Electric arc furnace
HCFCr   High carbon ferrochrome
HM   Hot metal
MR   Melt rate (kg/s)
SEC   Standard energy consumption (J/kg)

4.1 Melting Stock: Mixing Ratio and Environment

4.1.1 Melting Stock and Mixing Ratio

Melting stock and practice are important aspects of foundry operations. They have a direct bearing on the quality of casting. Melting furnaces are charged with melting stock, which consists of solid and/or liquid metal, alloying elements, flux, and slag-forming materials. It is important to decide the material-mixture ratio for a specific melting operation based on the casting requirement. For example, a typical material mixing ratio can be used for casting aluminum parts in a foundry: 40% aluminum ingot +50% aluminum scrap +10% other (alloying elements, flux, etc.).

The melting stock may be composed of commercially pure metals (such as aluminum, zinc, copper, etc.). Alloying elements are often added either in the melting furnace or in the ladle. In the case the melting temperatures of the alloying elements are sufficiently low, pure alloying elements are added to obtain the desired composition of the melt. On the other hand, if the melting temperatures are too high, master alloys may be used. These alloys consist of lower-melting-point alloys (Liebermann, 1993).

Care must be exercised to ensure that the specific gravity of the master alloy should not be too high, else segregation will occur in the casting.

Ingots and scrap metal are the main material constituents (around 80% of the mixing ratio) of melting stock used in a foundry. Prior to charging into the melting furnace, ingots and metal scraps (cans, scrap machinery parts, containers, or sidings) require cleaning (by removing dirt) followed by drying and preheating. The drying step helps to remove the moisture, limit the slag formation, and enhance the melting capacity of the furnace. In particular, preheating scrap metal is important for removing the paint, machining oil, and other contaminants.

4.1.2 Melting Stock and Environment

Melting stock must be environmentally friendly; particularly, it must be free from toxic elements, such as lead (Pb). Melting of scrap metal in foundries in low-income countries is crucial, especially in the cottage industry that melts scrap metal for making cookware. Very recently, Kuhangana and co-researchers (2024) have reported exposure to Pb among artisanal workers, and their families, involved in manufacturing cookware from scrap metal in Lubumbashi, DR Congo. They collected surface dust in the work-spaces, and blood and urine samples among workers, as well as residents living in the cookware workshops, and have reported high Pb concentrations (347 mg/kg) in the blood and urine samples. Resident children, from the cookware foundries, had higher urinary Pb contamination (6.2 µg/g creatinine) than adults (Kuhangana et al., 2024).

4.1.3   Flux and Slag-Forming Materials

A flux is a special blend of typically solid, inorganic compound that is designed for degassing, cleaning, grain refining, and/or chemistry/microstructure modification. Fluxes refine the molten metal by removing dissolved gases and impurities. They are usually added during melting, holding, or degassing of metal to furnaces, crucibles, ladles, or other vessels. Fluxes have several functions, depending on the metal being melted. For example, in the case of aluminum alloys, the following types of fluxes are used: (a) cover fluxes (to form a barrier to oxidation), (b) cleaning fluxes, (c) refining fluxes, and (d) drossing fluxes.

Besides aluminum alloys, fluxes are also used for melting gray cast iron; here a special flux (inoculant) is generally used to achieve the desirable microstructural control, preferably Type A graphite flakes in gray cast iron. Type A graphite flakes have a uniform distribution with a random orientation (Fig 4.1a); they reflect optimum mechanical behavior (American Foundry Society, 2000). This author has experience of failure analysis of an engine-block's cylinder liner (made of gray cast iron) that developed cracks due to incorrect graphite flakes morphology (the

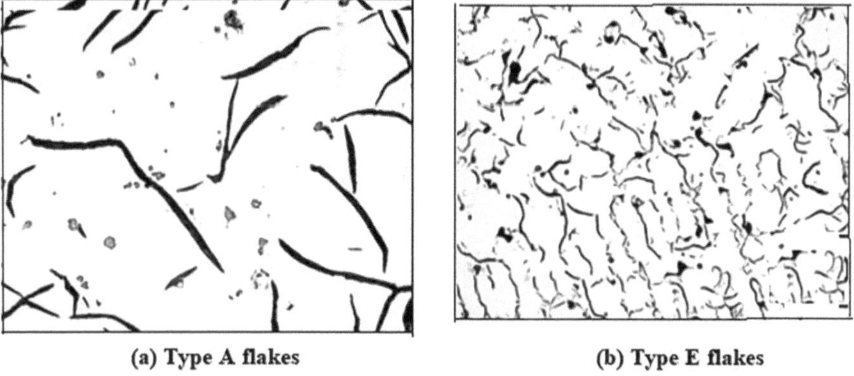

(a) Type A flakes   **(b) Type E flakes**

Fig. 4.1 Two (of the five) types of graphite flakes morphologies in gray cast iron: (**a**) Desirable Type A morphology. (**b**) Undesirable Type E morphology

Table 4.1 Capacities and capabilities of melting furnaces

Furnace	EAF	Induction furnace	Crucible furnace	Cupola
Metal melted	Steels	Ferrous and nonferrous	Nonferrous	Cast irons
Capacity (usual)	1–300 tons	1–60 tons	5–500 kg	1–100 tons/h

presence of Type E flakes) in its microstructure, which in turn was owing to the absence of the use of the inoculant during the melting practice (see Figs. 4.1a, b).

Slag forming materials are typically used for thermal insulation of molten metal. They serve the following purposes: (a) to protect the surface of the molten metal against atmospheric oxidation and contamination, (b) to refine the melt, and (c) to prevent heat loss. In order to provide thermal insulation by slag-forming material, special chemical compounds are covered onto the surface of the melt that combine with the metal to form a slag. In steel melting, the constituents of the slag include CaO, SiO_2, FeO, and MnO.

4.2 Melting Furnaces in Foundries

A foundry's melting furnace is equipment that is designed to overheat the metal stock to above its melting temperature—pouring temperature. The main function of a melting furnace is to melt metal stock so that it can be poured into molds to create various products. Foundry furnaces are designed to melt a wide range of metals, including cast iron, steel, aluminum, brass, bronze, and the like. The following types of furnaces are commonly used in foundries: (a) electric arc furnaces (EAF), (b) induction furnaces, (c) crucible furnaces, and (d) cupolas. The working principles of these furnaces (except the cupola) are explained in the following sections. The working principle of cupolas is discussed in the next chapter (Chap. 5). Table 4.1 presents the capacities of various melting furnaces used in foundries. The table also shows the capabilities of melting different types of metals in the furnaces.

4.3 Electric Arc Furnace

4.3.1 Construction and Working Principle

An electric arc furnace (*EAF*) is basically designed to melt and refine scrap steel in foundries. The capacities of EAFs are generally in the range of 1–300 tons (see Table 4.1). The arc furnace is essentially a steel shell with a spheroid bottom that is lined with refractory bricks. It has a tilting mechanism and lip for pouring. The *EAF* operation involves the use of large graphite electrodes that create electric arcs to melt the metal. The graphite electrodes are lowered into the furnace to create an electric arc between the metal stock and the electrodes (see Fig. 4.2).

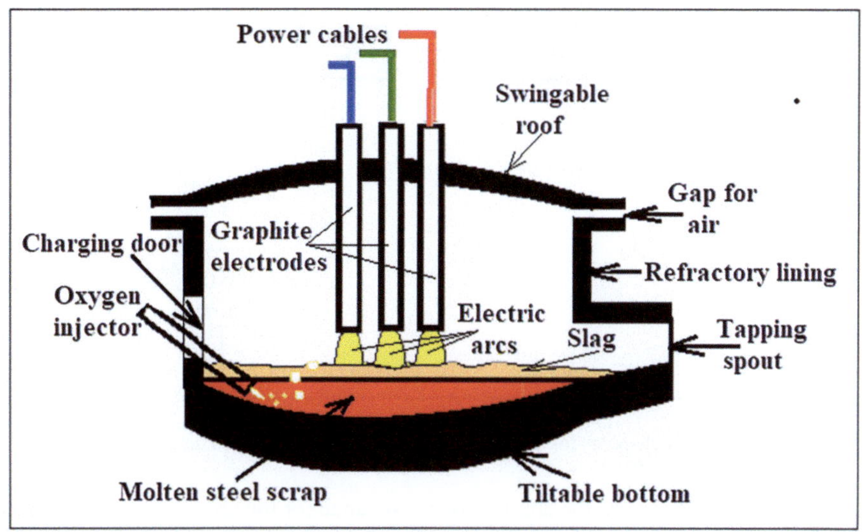

Fig. 4.2 Schematic of an electric arc furnace for melting steel scrap

Firstly, the swing-able roof is moved aside and the scrap steel is charged into the furnace by use of an overhead crane. A furnace transformer is connected to the electrodes through power cables, which supply high amperage current (around 100 kA). The transformer secondary voltage, V_s, is varied from 120 to 800 volts, depending on the size of the furnace and the operation being performed. In order to produce a ton of steel in an EAF, a minimum of 300 kWh (approx.) heat energy is required (see Example 4.1). The intense heat generated by the arcs melts the steel scrap, which then is refined by the addition of slag-forming materials (lime or fluorspar). Additionally, coke addition and oxygen-lancing is done to speed up the refining process. Once the desired level of purity has been achieved, the molten metal is tapped from the arc furnace and poured into a ladle for subsequent pouring to casting molds. A distinct advantage of the electric arc furnace is that it can be used to recycle scrap steel. Hence, EAF is more environmentally friendly as compared to furnaces that rely on fossil fuels.

4.3.2 Electrical Engineering Analysis

The working principle of an electric arc furnace (EAF) is based on the conversion of electrical energy into heat energy. The EAF circuit elements, for the peak power transferred from V_s to the load, are shown in Fig. 4.3. The main EAF circuit elements include: the transformer line's secondary voltage V_s, the electrode current I_s (assumed to be sinusoidal), the source reactance X_s, the source resistance R_s, and the load resistance X_L.

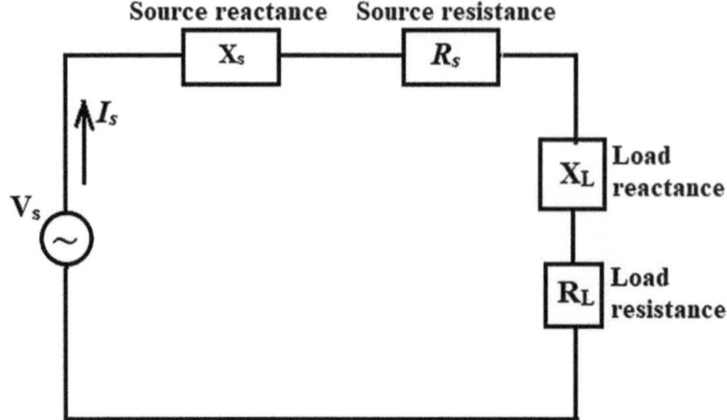

Fig. 4.3 EAF circuit elements for the peak power transferred from V_s to the load

For the peak power to the arc, the load resistance (or the arc resistance) is related to the source resistance, source reactance, and the load reactance by the following formula (Martell-Chávez et al., 2013):

$$R_L = \sqrt{R_s^2 + \left(X_s + X_L\right)^2} \tag{4.1}$$

where R_L is the load resistance (mΩ), R_s is the source resistance (mΩ), X_s is the source reactance (mΩ), and X_L is the load reactance (mΩ) (see Example 4.2). Once the load resistance has been computed, the peak power delivered to the arc, P_{peak}, can be determined by:

$$P_{peak} = \frac{\left|V_s\right|^2}{2\left(R_s + R_L\right)} \tag{4.2}$$

where P_{peak} is the peak power (W), V_s is the transformer's secondary voltage (V), R_s is the source resistance (Ω), and R_L is the load resistance (Ω) (see Example 4.3).

4.3.3 Furnace Capacity Calculation

The electric arc furnace, in general, supplies molten steel to a billet/slab continuous casting (CC) machine (see Chap. 1, Fig. 1.6). The capacity of an EAF depends on a number of factors, including the time demand of billet caster, t, the billet size to be cast, the casting speed of the CC machine, v, the density of steel, and the number of strands in the CC machine, n. The EAF capacity can be calculated by the following formula:

$$\text{Capacity}_{EAF} = t \bullet v \bullet \rho_{steel} \bullet n \bullet \left(\text{billet c.s.a}\right) \tag{4.3}$$

where capacity$_{EAF}$ is the capacity of the electric arc furnace (tons/heat), t is the time taken by billet caster (min), v is the average casting speed of the CC machine (m/min), n is the number of strands in the CC plant, and the billet cross-sectional area is in m × m units (see Example 4.4).

4.3.4   Charge Calculations in EAF

4.3.4.1   Melting Stock in EAF

It is learnt in the preceding section that scrap steel (cold) is the main melting stock in an EAF. In general, in order to produce one ton of molten steel (carbon steel) in an EAF, the mass of carbon-steel scrap required is in the range of 1030–1045 kg. Although scrap steel is the main charging material in an EAF, there are other materials that are usually added through the charging door; these charging materials include: lime, fluorspar, coal, oxygen, hot metal (HM), ferroalloys, pure alloying elements, etc. The additions of ferroalloys and pure elements are carried out only in the manufacture of alloy steels. The charge calculation for an EAF may be done based on a material balance model for the melting period. Recently, Petrov has reported the material balance modeling during the melting period for bearing steel—a type of tool steel (Petrov, 2023).

4.3.4.2   Scrap Calculation

Steel scrap can be charged either by an overhead crane or by use of buckets. In the case scrap is charged by buckets, the amount of scrap to be charged into an EAF can be determined by the following formula:

$$m_{\text{scrap}} = \frac{\text{Capacity}_{\text{EAF}}}{Fe_{\text{scrap}} \bullet n_{\text{bucket}}} \tag{4.4}$$

where m_{scrap} is the mass of scrap to be charged by use of a bucket (tons), Fe_{scrap} is the Fe content in the scrap (wt.%), and n_{bucket} is the number of bucket provided for charging (see Example 4.5).

4.3.4.3   Hot Metal Charging: Benefits to Economy and Environment

The charging of hot metal (HM), in EAF practice, has shown to be highly beneficial to the energy economy and environment. Recently, Irawan and co-researchers (2022) have reported that the electric arc power consumption can be reduced down to 72.9 MWh from 87.4 MWh (16% more energy efficient) while at the same time increasing the HM charging temperature at the endpoint of the duct de-dusting system up to 900 °C from 540 °C (Irawan et al., 2022).

Additionally, recent research has shown a significant reduction in carbon emissions from EAF, by HM mixing with scarp. It has been reported that the carbon emission of power (per ton of molten steel) decreased from 143 kg to 20.4 kg as the HM/scrap ratio was increased from 0 to 70% (Yang et al., 2016). The carbon emission of power per ton of molten steel can be determined by the following formula:

$$M_{EAF}^{i} = \frac{\text{Molecular mass of } CO_2}{\text{Atomic mass of carbon}} \times Q_{EE}^{i} \bullet C_{Scoal} \tag{4.5}$$

or

$$M_{EAF}^{i} = \frac{44}{12} \times Q_{EE}^{i} \bullet C_{Scoal} \tag{4.6}$$

where M_{EAF}^{i} is the carbon emission of power per ton of molten steel (kg), i is the hot metal (HM) ratio, Q_{EE}^{i} is the electrical energy consumption per ton of molten steel (J), and C_{Scoal} is the carbon content of standard coal (C_{Scoal}= 25.8 kg/GJ) (see Examples 4.6 and 4.7).

4.3.4.4   Alloy Steel Manufacture in EAF

Alloy steel is the steel made by adding one or more alloying elements to carbon steel. The alloying elements include manganese (Mn), silicon (Si), nickel (Ni), chromium (Cr), molybdenum (Mo), vanadium (V), etc. Alloy steel castings have improved strength, hardness, toughness, corrosion resistance, wear resistance, and the like. Alloying can be done either by adding pure elements or by adding ferroalloys. In the case a pure alloying element is added to the steel-holding ladle, the mass of the element required to be added can be determined by:

$$m_{additive} = \frac{\Delta\%X \times m_{ladle}}{100} \tag{4.7}$$

where $m_{additive}$ is the mass of the element required to be added (kg), m_{ladle} is the mass of the molten steel in the ladle (kg), and $\Delta\%X$ is the required increase in the *wt.%* of the alloying element X ($\Delta\%X = \%X_{aim} - \%X_{actual}$) (see Example 4.8).

In industrial practice, owing to the economy of the technique, alloying is usually done by the addition of ferroalloys. In such a case, it is important to consider "% *recovery*" (wt.% of the element that actually increases the liquid steel composition rather than being lost to the slag) in the calculation. The mass of the ferroalloy ($m_{Fe-alloy}$) required to be added can be determined by (Huda, 2020):

$$m_{Fe-alloy} = \frac{100 \times \Delta\%X \times m_{ladle}}{\%X \text{ in master ferroalloy} \times \% \text{recovery of } X} \tag{4.8}$$

The significance of Eq. 4.8 is illustrated in Examples 4.9, 4.10, 4.11, 4.12, and 4.13.

Some high-carbon ferroalloys can cause an increase in the carbon (C) content of alloy steels. In such cases, the amount of carbon pick up by steel (Δ wt.% C) can be determined by:

$$\Delta\%C = \frac{m_{\text{ferroalloy}} \times \%C \text{ in ferroalloy} \times \% \text{recovery of C}}{100 \times m_{\text{ladle}}} \tag{4.9}$$

The significance of Eq. 4.9 is illustrated in Example 4.14.

4.4   Induction Furnace

4.4.1   Working Principle

An induction furnace is an electrical-energy based foundry furnace. It is so named because the heat energy is supplied by induction heating of metal. An induction furnace is capable of melting both ferrous and nonferrous scrap metals and alloys. The capacities of induction furnaces range from 1 ton to 60 tons. The operation of an induction furnace involves the passing of alternating current (a.c) through copper coils to develop a magnetic field in metal. The induced (eddy) current in the molten metal causes agitation and rapid heating of solid metal stock resulting in melting (see Fig. 4.4). Induction furnaces are capable of producing molten metals/alloys of

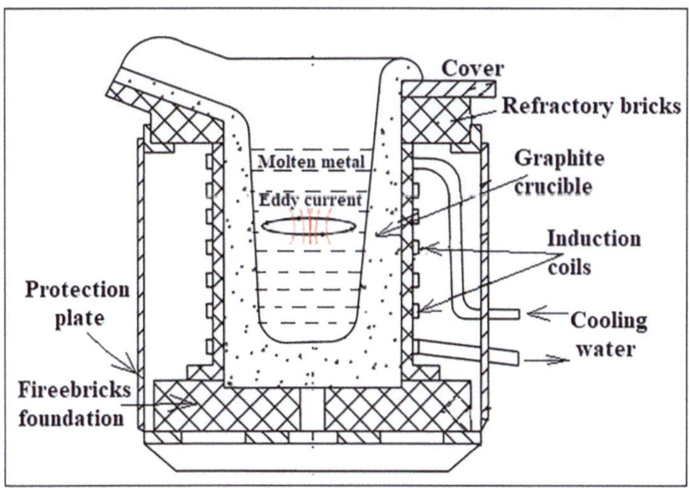

Fig. 4.4 Schematic of an induction furnace

high quality and purity. The commonly melted metals, in foundries, include copper, zinc, steels, cast irons, aluminum alloys, and the like.

4.4.2　Analysis for Power, Energy, and Cost

The power consumed in melting a specific metal, in an induction furnace, strongly depends on the standard energy consumption (*SEC*) of the metal, which is the energy consumed at standard kW/kg ratio, generally expressed in kWh/ton. The standard energy consumption (in kWh/ton) data for various metals/alloys is presented in Table 4.2.

The gross power of an induction furnace can be determined by the following formula:

$$P_g = \frac{(MR) \bullet (SEC)}{\eta} \tag{4.10}$$

where P_g is the furnace's gross power (W), MR is the melt rate (kg/s), SEC is the standard energy consumption (J/kg) for melting the particular metal, and η is the furnace efficiency (see Example 4.15). The time required to melt a given mass of a specific metal, in the induction furnace, can be determined by:

$$\text{Time required} = \frac{SEC}{P_g} \times \text{Melt mass} \tag{4.11}$$

The significance of Eq. 4.11 is illustrated in Example 4.16.

The monthly cost on electrical energy consumption can be determined by:

$$\text{Monthly cost on electricity} = \$\,x\,/\,kWh\,\times\,SEC\,kWh\,/\,ton \\ \times \text{ monthly tonnage melted} \tag{4.12}$$

The significance of Eq. 4.12 is illustrated in Example 4.17.

Table 4.2　Standard energy consumption (in kWh/ton) for various metals/alloys

Material	Cast iron (ingot)	Aluminum (light scrap)	MS/SS	SG iron
SEC (kWh/ton)	560	600–625	600–650	550–600

MS = mild steel, SS = stainless steel, SG = spheroidal graphitic cast iron

4.5   Crucible Furnace

4.5.1   Working Principle

A crucible furnace is a foundry furnace that is primarily used for melting nonferrous metals with a low melting temperature (e.g. aluminum, brass, bronze, etc.). It is so named because the container that the metals and additives are placed in is referred to as the "crucible," which is made of graphite—a heat-resistant material with good thermal conductivity. The charge is heated via conduction of heat through the walls of the crucible. The heating fuel is generally coke, oil, or gas (see Fig. 4.5).

A crucible furnace is commonly used where small batches of low melting-point alloys (e.g., zinc, brass, aluminum alloys, etc.) are required. In order to operate the furnace, the crucible is placed inside the furnace chamber and heated by a burner, until the metal inside it melts (see Fig. 4.5). Once the metal has been melted, it is poured into casting molds. The crucible is generally removed from the furnace chamber by use of tongs or other suitable tools.

4.5.2   Types of Crucible Furnaces

The capacities of crucible furnaces, in general, are in the range of 5–500 kg. A crucible may be a pit furnace or a furnace built on the ground level. Based on the method of heating, there are three types of crucible furnaces: (a) coke fired, (b) oil fired, and (c) gas fired. Crucible furnaces are typically classified according to the method of removing the metal from the crucible, as follows: (1) tilting furnace, (2) lift-out furnace, and (3) bale-out furnace. *Tilting furnace* involves transferring the molten metal to the mold or ladle by mechanically tilting the crucible and furnace body. *Lift-out furnace* is the one in which the crucible and molten metal are removed from the furnace body for direct pouring into the mold (see Fig. 4.6). In a *bale-out furnace*, the molten metal is ladled from the crucible to the mold.

Fig. 4.5 Schematic of gas-fired crucible furnace

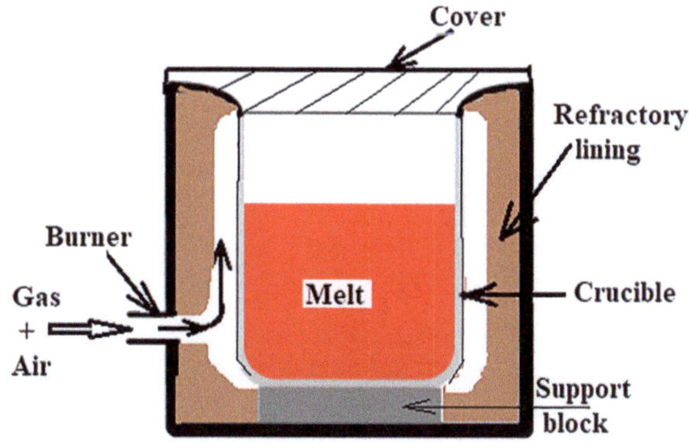

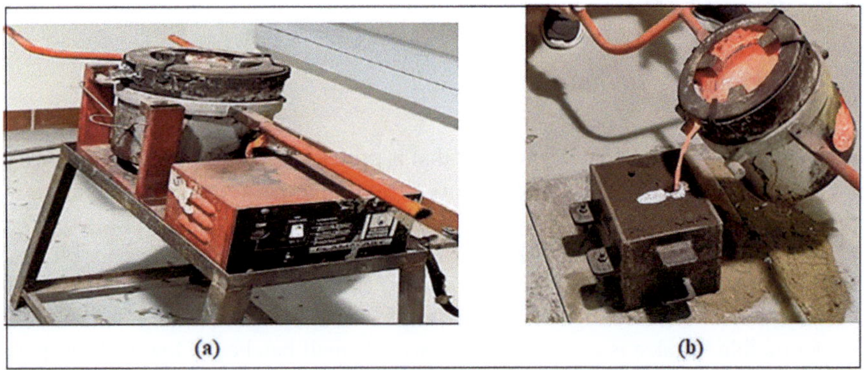

Fig. 4.6 A lift-out crucible furnace (**a**). Pouring of molten metal into a sand mold (**b**)

4.5.3 *Furnace Design*

4.5.3.1 Design of Crucible Pot

The essential components that are used in the construction of a crucible furnace include crucible- pot, air blower, fire-bricks, fire nozzle, furnace drum, and the cover. The crucible pot of the furnace can be designed by determining the volume of the crucible by:

$$V_c = \frac{M_m}{\rho_m} \tag{4.13}$$

where V_c is the volume of crucible (m³), M_m is the mass of the metal in the crucible (kg), and ρ_m is the liquid density of the metal (kg/m³). The height of the crucible can be calculated by:

$$V_c = \frac{\pi}{4} \bullet D^2 \bullet h_c \tag{4.14}$$

$$h_c = \frac{4}{\pi} \bullet \frac{V_c}{D^2} = 1.273 \times \frac{V_c}{D^2} \tag{4.15}$$

where h_c is the height of the crucible (m), and D is the internal diameter of the crucible (m) (see Example 4.18). The internal circumference (C_c) of the crucible pot can be calculated by:

$$C_c = \pi \bullet D_c \tag{4.16}$$

Fig. 4.7 Refractory bricks arrangement in a crucible furnace: (**a**) seating bricks, (**b**) standing bricks

4.5.3.2 Design of Refractory Bricks

The crucible furnace is lined with refractory bricks that are laid between the crucible-pot and the drum. The linings are done by arranging the seating bricks at the bottom, and the standing bricks at the remaining part of the furnace (see Fig. 4.7). The volume of a brick V_b can be calculated by:

$$V_b = l_b \bullet w_b \bullet t_b \tag{4.17}$$

where l_b, w_b, and t_b are the length, width, and thickness of the brick, respectively.

The internal height of the furnace, $h_{i(\text{furnace})}$, can be determined by (Joseph et al., 2016):

$$h_{i(\text{furnace})} = h_c + \frac{3}{2} l_b \tag{4.18}$$

where h_c is the crucible height and l_b is the brick's length. The external height of the furnace $h_{e(\text{furnace})}$ can be determined by:

$$h_{e(\text{furnace})} = h_{i(\text{furnace})} + 2 \bullet t_b + \text{steel} - \text{sheet thickness} \tag{4.19}$$

The significance of Eqs. 4.18 and 4.19 is illustrated in Examples 4.19 and 4.20.

In order to design the refractory bricks for the crucible furnace, it is important to determine the number seating bricks as well as the number of standing bricks (see Fig. 4.6). The number of seating bricks at the bottom (Fig. 4.7a) can be determined by:

$$\text{No. of seating bricks} = \frac{C_f}{l_b} \tag{4.20}$$

where C_f is the internal circumference of the furnace (see Example 4.21). The number of standing bricks (Fig. 4.7b) can be determined by:

$$\text{No. of standing bricks} = \frac{C_f}{w_b} \tag{4.21}$$

The significance of Eqs. 4.20 and 4.21 is illustrated in Examples 4.22 and 4.23. The number of stacking layers of bricks can be determined by:

$$\text{No. of stacking layers of bricks} = \frac{h_{i(\text{furnace})} - 2 \bullet t_b}{l_b} \tag{4.22}$$

The significance of Eq. 4.22 is illustrated in Example 4.24.

4.6   Worked Examples in Melting Furnaces

Example 4.1 Determining the Total Energy Required to Melt and Superheat Steel in EAF

The thermal data for scrap steel is given in Table 4.3. Determine the minimum amount of heat energy required to raise the temperature of a metric ton of steel scrap from the ambient temperature (20 °C) to a temperature of 1520 °C in an electric arc furnace.

Solution

With reference to Table 4.3,

$m = 1$ metric ton $= 1000$ kg, $T_m = 1450$ °C, $C_s = 420$ J/kg·°C, $C_{liq} = 750$ J/kg °C, $\Delta H_f = 268$ kJ/kg.

By using Eq. 2.6,

$Q = m\ [C_s{\cdot}(T_m\text{-}T_o) + H_f + C_L \cdot (T_p\text{-} T_m)]$.

$Q = 1000 \times \{[420 \times (1450\text{--}20)] + 268{,}000 + [750 \times (1520\text{--}1450)]\}$.

$Q = 1000 \times [(420 \times 1430) + 268{,}000 + (750 \times 70)] = 1000 \times (600{,}600 + 268{,}000 + 52{,}500)$.

Heat energy required $= Q = 921{,}100{,}000$ J $= 921{,}100{,}000 \times 2.778 \times 10^{-7}$ kWh $= 255.88$ kWh.

Table 4.3 Thermal data for Example 4.1

Melting temperature	Specific heat of solid steel	Specific heat of liq. Steel	ΔH_f
1450 °C	420 J/kg °C	750 J/kg °C	268 kJ/kg

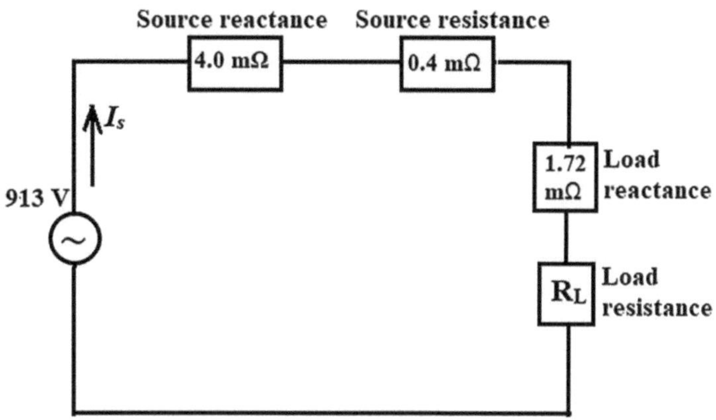

Fig. 4.8 The EAF circuit elements for Example 4.2

Example 4.2 Determining the EAF's Load Resistance for the Peak Power to the Arc

The data for the EAF circuit elements, for the peak power transferred from V_s to the load, are shown in Fig. 4.8. Determine the arc resistance for the peak power delivered to the load.

Solution

$R_s = 0.40$ mΩ, $X_s = 4.0$ mΩ, $X_L = 1.72$ mΩ, $R_L =?$

By using Eq. 4.1,

$$R_L = \sqrt{R_s^2 + (X_s + X_L)^2} = \sqrt{0.4^2 + (4.0 + 1.72)^2} = \sqrt{0.16 + 32.7184} = 5.73 \text{ m}\Omega$$

The arc resistance for the peak power to the load $= R_L = 5.73$ mΩ $= 5.73 \times 10^{-3}$ Ω

Example 4.3 Determining the Peak Power Delivered to the Arc in EAF

By using the data in Example 4.2, determine the peak power delivered to the arc in the EAF.

Solution

$V_s = 913$ V, $R_s = 0.40 \times 10^{-3}$ Ω, $R_L = 5.73 \times 10^{-3}$ Ω, $P_{peak} =?$

By using Eq. 4.2,

$$P_{peak} = \frac{|V_s|^2}{2(R_s + R_L)} = \frac{|913|^2}{2(0.40 \times 10^{-3} + 5.73 \times 10^{-3})} = 68213.5 \times 10^3 \text{ W} = 68.213 \text{ MW}.$$

The peak power delivered to the arc in the EAF $= 68.213$ MW.

Example 4.4 Determining the Capacity of an EAF Linked with a Billet C-C Machine

An EAF has to supply molten steel to a 4-strand CC plant that casts billets of cross-sectional area (140 mm × 140 mm) in an hour at an average casting speed of 2250 mm/min. Determine the capacity of the furnace.

Solution

Billet c.s.a = 140 mm × 140 mm = 0.14 m × 0.14 m, t = 1 h = 60 min,
   n = 4, ρ_{steel} = 7.8 × 10^3 kg/m^3 = 7.8 tons/m^3, v = 2250 mm/min = 2.25 m/min.
   By using Eq. 4.3,

Capacity$_{EAF}$ = t•v•ρ_{steel}•n•(billet c.s.a) = 60 × 2.25 × 7.8 × 4 × 0.14 × 0.14
Capacity$_{EAF}$ = 82.555 tons/heat.

Example 4.5 Determining the Mass of Scrap to Be Added to an EAF

A foundry manager has purchased steel scrap with Fe content of 95%. Four (4) buckets are provided to charge the scrap to an EAF with a capacity of 80 tons/heat. What amount of scrap must be charged by use of a bucket?

Solution

Fe_{scrap} = 95% = 0.95, n_{bucket} = 4, Capacity$_{EAF}$ = 80 tons/heat.
   By using Eq. 4.4,

$$m_{scrap} = \frac{\text{Capacity}_{EAF}}{Fe_{scrap} \cdot n_{bucket}} = \frac{80}{0.95 \times 4} = 21.052 \text{ tons.}$$

It means that at least 21.052 tons scrap must be charged by use of a bucket.

Example 4.6 Determining the Carbon Emission of Power Per Ton of Molten Steel in EAF

The carbon emission of power per ton of molten steel with no addition of hot metal in an EAF is 143 kg. The electrical-energy consumption with use of hot metal ratio of 30% was experimentally determined to be 275 kWh/ton. Determine the carbon emission of the energy per ton of molten steel with hot metal ratio of 30%.

Solution

Q_{EE}^{30} = 275 kWh/ton = 275 × 3.6 × 10^6 J = 990 × 10^6 J,
   C_{Scoal} = 25.8 kg/GJ = 25.8 × 10^{-9} kg/J, M_{EAF}^{0} = 143 kg, M_{EAF}^{30} =?
   By using Eq. 4.6,

$$M_{EAF}^{30} = \frac{44}{12} \times Q_{EE}^{30} \cdot C_{Scoal} = \frac{44}{12} \times 990 \times 10^6 \times 25.8 \times 10^{-9} = 93.654 \text{ kg.}$$

Carbon emission of the energy per ton of molten steel with hot metal ratio of 30% = 93.654 kg.

Example 4.7 Determining the Percent Reduction in Carbon Emission in EAF

By using the data in Example 4.6, determine the percentage reduction in carbon emission of power due to the use of 30% hot-metal ratio.

Solution

M_{EAF}^{0} = 143 kg, M_{EAF}^{30} = 93.654 kg.

$$\% \text{ Reduction in carbon emission} = \frac{M_{EAF}^{0} - M_{EAF}^{30}}{M_{EAF}^{0}} \times 100 = \frac{143 - 93.654}{143} \times 100 = 34.5.$$

Reduction in carbon emission due to the use of 30% hot-metal ratio = 34.5%.

Example 4.8 Calculating the Mass of an Alloying Element Required to Be Added

Spectrometry analysis of steel from a 15-tons-ladle shows 0.32 wt.% Mn. Calculate the mass of manganese (Mn) required to be added to achieve an aim composition of 1.50 wt.% Mn.

Solution

m_{ladle} = 15 tons = 15,000 kg, wt.% Mn_{actual} = 0.32, wt.%Mn_{aim}= 1.50, Δ % Mn = % Mn_{aim}—% Mn_{actual}= 1.50–0.32 = 1.18.

By using Eq. 4.7,

$$m_{additive} = \frac{\Delta\%Mn \times m_{ladle}}{100} = \frac{1.18 \times 15,000}{100} = 177 \text{ kg.}$$

The mass of the elemental manganese required to be added = 177 kg.

Example 4.9 Determining Mass of a Ferrosilicon Required to Be Added to Produce Alloy Steel

It is required to produce 15-tons molten steel in EAF. The wt.% of various alloying elements and their recoveries in the corresponding ferroalloys is given in Table 4.4.

The aim wt.% and the actual wt.% of the alloying elements in the ladle steel are presented in Table 4.5.

Calculate the mass of ferrosilicon required to be added to the ladle to achieve the aim composition.

Solution

m_{ladle} = 15 ton = 15,000 kg, Δ % Si = % Si_{aim} − %Si_{actual} = 0.25–0.05 = 0.20% recovery = 95, % Si in ferrosilicon = 75, $m_{ferrosilicon}$ =?

By using Equ. 4.8,

$$m_{ferrosilicon} = \frac{100 \times \Delta\%Si \times m_{ladle}}{\%Si \text{ in ferrosilicon} \times \%\text{recovery Si}} = \frac{100 \times 0.20 \times 15,000}{75 \times 95} = 42.1 \text{ kg.}$$

The required mass of ferrosilicon = 42.1 kg.

Table 4.4 Wt.% alloying elements and recoveries in ferroalloys

–	Si in ferrosilicon	Mn in ferromanganese	Ni in ferronickel	Cr in ferrochrome	Mo in ferromolybdenum
Wt.%	75	75	80	70	62
% recovery	95	90	85	78	80

Table 4.5 Alloying elemental data for alloy steel manufacture

–	Silicon (Si)	Manganese (Mn)	Nickel (Ni)	Chromium (Cr)	Moly (Mo)
Aim %	0.25	0.7	4	1	0.4
Actual %	0.05	0.1	0.8	0.2	0.08

Example 4.10 Determining Mass of a Ferromanganese to Be Added to Produce Alloy Steel

By using the data in Example 4.9, determine the mass of ferromanganese required to be added to the ladle to achieve the aim composition.

Solution

m_{ladle} = 15 ton = 15,000 kg, Δ % Mn = % Mn_{aim} − % Mn_{actual} = 0.7–0.1 = 0.6% recovery = 90, % Mn in ferromanganese = 75, $m_{ferromanganese}$ =?

$$m_{ferromanganese} = \frac{100 \times \Delta\%Mn \times m_{ladle}}{\%Mn\,in\,ferromanganese \times \%recovery\,Mn} = \frac{100 \times 0.6 \times 15,000}{75 \times 90} =$$

133.33 kg.

The required mass of ferromanganese = 133.33 kg.

Example 4.11 Determining Mass of a Ferronickel Required to Be Added to Produce Alloy Steel

By using the data in Example 4.10, determine the mass of ferronickel required to be added to the ladle to achieve the aim composition.

Solution

m_{ladle} = 15 ton = 15,000 kg, Δ % Ni = % Ni_{aim} − % Ni_{actual} = 4.0–0.8 = 3.2% recovery = 85, % Ni in ferronickel = 80, $m_{ferronickel}$ =?

$$m_{ferronickel} = \frac{100 \times \Delta\%Ni \times m_{ladle}}{\%Ni\,in\,ferronickel \times \%recovery\,Ni} = \frac{100 \times 3.2 \times 15,000}{80 \times 85} = 705.88\ kg.$$

The required mass of ferronickel = 705.88 kg.

Example 4.12 Determining Mass of a Ferrochrome to Be Added to Produce Alloy Steel

By using the data in Example 4.11, determine the mass of ferrochrome required to be added to the ladle to achieve the aim composition.

Solution

m_{ladle} = 15 ton = 15,000 kg, Δ % Cr = % Cr_{aim} − % Cr_{actual} = 1.0–0.2 = 0.8% recovery = 78, % Cr in ferrochrome = 70, $m_{ferrochrome}$ = ?

$$m_{ferrochrome} = \frac{100 \times \Delta\%Cr \times m_{ladle}}{\%Cr\,in\,ferrochrome \times \%recovery\,Cr} = \frac{100 \times 0.8 \times 15,000}{70 \times 78} = 219.78\ kg.$$

The required mass of ferrochrome = 219.78 kg.

Example 4.13 Determining Mass of a Ferromolybdenum to Be Added to Produce Alloy Steel

By using the data in Example 4.12, determine the mass of ferromolybdenum required to be added to the ladle to achieve the aim composition.

Solution

m_{ladle} = 15 ton = 15,000 kg, Δ % Mo = % Mo_{aim} − % Mo_{actual} = 0.4–0.08 = 0.32% recovery = 80, % Mo in ferromolybdenum = 62, $m_{ferromolybdenum}$ =?

$$m_{\text{ferromolybdenum}} = \frac{100 \times \Delta\% Mo \times m_{\text{ladle}}}{\% Mo \text{ in ferromoly} \times \% \text{recovery } Mo} = \frac{100 \times 0.32 \times 15,000}{62 \times 80} =$$

96.77 kg.

The required mass of ferromolybdenum = 96.77 kg.

Example 4.14 Determining the wt.% Carbon Pickup Due to Ferroalloy Addition

It is required to produce 12 tons of molten alloy steel in an EAF. Determine the wt.% of carbon pickup due to the addition of 800 kg high-carbon ferrochrome (HCFeCr) in the ladle. The ferrochrome contains 6.5 wt.% carbon with 80% recovery.

Solution

m_{ladle} = 12,000 kg, m_{FeCr} = 800 kg, wt.% C in FeCr = 6.5, % recovery of carbon = 80.

By using Eq. 4.9,

$$\Delta\% \text{ C} = \frac{m_{\text{FeCr}} \times \text{wt.}\% C \text{ in FeCr} \times \% \text{recovery of } C}{100 \times m_{\text{ladle}}} = \frac{800 \times 6.5 \times 80}{100 \times 12000} = 0.35.$$

The amount of carbon pick-up = 0.35 wt.%.

Example 4.15 Determining the Gross Power of an Induction Furnace

The rate of melting stainless steel, in an induction furnace, is 520 kg/h. The furnace efficiency is 88%. Determine the gross power of the induction furnace. Hint: refer to the data in Table 4.2.

Solution

$$\text{Melting rate} = \text{MR} = \frac{520\,\text{kg}}{1\,\text{h}} = \frac{520\,\text{kg}}{60 \times 60\,\text{s}} = 0.14444 \text{ kg/s}, \eta = 88\% = 0.88.$$

With reference to the data in Table 4.2, SEC for steel = 625 kWh/ton.

$$\text{SEC} = \frac{625\,\text{kWh}}{1\,\text{ton}} = \frac{625 \times 1000\,\text{W} \times 3600\,\text{s}}{1000\,\text{kg}} = 2,250,000 \text{ W} \cdot \text{s/kg}.$$

By using Eq. 4.10,

$$P_g = \frac{(\text{MR}) \cdot (\text{SEC})}{\eta} = \frac{0.1444\,\text{kg/s} \times 2250000\,\text{W} \cdot \text{s/kg}}{0.88} =$$

369204.54 W = 369.204 kW.

The gross power of the induction furnace $\cong$ 370 kW.

Example 4.16 Determining the Time Required to Melt 1 Ton of Metal in Induction Furnace

By using the data in Example 4.15, determine the time required to melt 1 ton of cast iron (ingot) in the same induction furnace.

Solution

P_g = 370 kW = 370,000 W, Melt mass = 1 ton = 1000 kg,

$$\text{By reference to Table 4.2, SEC} = 560\,\text{kWh/ton} = \frac{560\,\text{kWh}}{1\,\text{ton}} = \frac{560 \times 1000\,\text{W} \times 3600\,s}{1000\,\text{kg}}$$

= 2,016,000 W•s/kg.

By using Eq. 4.11,

$$\text{Time required} = \frac{\text{SEC}}{P_g} \times \text{Melt mass} = \frac{2016000}{370000} \times 1000 = 5448.6\ s = 90.81\ \text{min}$$

$\cong 91$ min.

Example 4.17 Determining the Monthly Electrical Cost for Melting in Induction Furnace

A foundry melts 3 tons/day light scrap of aluminum in an induction furnace. Determine the monthly cost on electrical energy, if the unit price is \$ 0.11/kWh. Assume 1 month = 30 days.

Solution

With reference to Table 4.2, the $(\text{SEC})_{av}$ for *Al* light scrap $= \dfrac{600 + 625}{2} =$ 612.50 kWh/ton.

Monthly tonnage melted = 3 × 30 = 90 tons.

By using Eq. 4.12,

Monthly cost on electricity = \$ 0.11/kWh × 612.5 kWh/ton × 90 tons = \$ 6063.75.

Example 4.18 Designing a Crucible Furnace by Determining Its Crucible's Height

It is required to melt 70 kg aluminum in a crucible furnace with an internal diameter of 0.24 m. Design the crucible by determining its height. The density of liq. aluminum = 2373 kg/m³.

Solution

M_m = 70 kg, ρ_m = 2373 kg/m³, D = 0.24 m, h_c =?

By using Eqs. 4.13 and 4.15,

$$V_c = \frac{M_m}{\rho_m} = \frac{70}{2373} = 0.0295\ \text{m}^3.$$

$$h_c = 1.273 \times \frac{V_c}{D^2} = 1.273 \times \frac{0.0295}{0.24^2} = \frac{0.03755}{0.0576} = 0.652\ \text{m}.$$

The crucible height = h_c = 0.652 m = 652 mm.

Example 4.19 Determining the Internal Height of a Crucible Furnace

By using the data in Example 4.18, determine the internal height of the crucible furnace. The dimensions of the refractory-brick are 250 mm × 126 mm × 70 mm.

Solution

Data: l_b = 250 mm, w_b = 120 mm, t_b = 70 mm, h_c = 652 mm.

By using Eq. 4.18,

$$h_{i(\text{furnace})} = h_c + \frac{3}{2}l_b = 652 + \frac{3}{2} \times 250 = 1027 \text{ mm}.$$

The internal height of the furnace $= h_{i(\text{furnace})} = 1027$ mm $= 1.027$ m.

Example 4.20 Determining the External Height of a Crucible Furnace
By using the data in Example 4.19, determine the external height of the crucible furnace. The thickness of the steel-sheet of the drum is 6 mm.

Solution
Data: $h_{i(\text{furnace})} = 1027$ mm, $t_b = 70$ mm.
  By using Eq. 4.19,

$$h_{e(\text{furnace})} = h_{i(\text{furnace})} + 2 \cdot t_b + \text{steel-sheet thickness} = 1027 + 140 + 6 = 1173 \text{ mm}.$$

The external height of the furnace $= h_{e(\text{furnace})} = 1173$ mm $= 1.173$ m.

Example 4.21 Determining the Number of Seating Bricks Arranged in Double Layers
By using the data in Example 4.19, determine the number of seating bricks arranged in double layers at the bottom. The internal diameter of a crucible furnace is 900 mm.

Solution
D = 900 mm, l_b= 250 mm.

Internal circumference of the furnace $= C_f = \pi \cdot D_f = \pi \times 900 = 2827$ mm.

  By using Eq. 4.20,

$$\text{No. of seating bricks} = \frac{C_f}{l_b} = \frac{2827}{250} = 11.$$

Since the bricks are arranged in double layers, the number of seating bricks $= 2 \times 11 = 22$.

Example 4.22 Determining the Number of Standing Bricks Arranged in Double Layers
By using the data in Examples 4.19, 4.20 and 4.21, determine the number of standing bricks arranged in double layers.

Solution
Data: $w_b = 126$ mm, $C_f = 2827$ mm.
  By sung Eq. 4.21,

$$\text{No. of standing bricks} = \frac{C_f}{w_b} = \frac{2827}{126} = 22.4 = 22.$$

Since the bricks are arranged in double layers, the number of standing bricks $= 2 \times 22 = 44$.

Example 4.23 Determining the Stacking Layers of Bricks in a Crucible Furnace
By using the data in Examples 4.19, 4.20, 4.21, and 4.22, determine the number of stacking layers of bricks.

Solution

Data: $h_{i(\text{furnace})} = 1027$ mm, $t_b = 70$ mm, $l_b = 250$ mm.

No. of stacking layers of bricks $= \dfrac{h_{i(\text{furnace})} - 2 \bullet t_b}{l_b} = \dfrac{1027 - 2 \times 70}{250} = 3.5 = 3.$

Example 4.24 Determining the Total Number of Standing Bricks to Construct the Furnace

By using the data in Examples 4.19, 4.20, 4.21, 4.22, and 4.23, determine the total number of standing bricks needed to construct the crucible furnace.

Solution

No. of stacking layers of bricks = 3.

No. of standing bricks = 44.

The total number of standing bricks needed to construct the crucible furnace = 3 × 44 = 132.

Questions and Problems

4.1. Encircle the most appropriate answers for each of the following statements:

(1) Which one is the most commonly used furnace for melting cast irons?
       (a) EAF, (b) cupola, (c) induction furnace, (d) crucible furnace
(2) What is the usual capacity range of electric arc furnaces?
       (a) 1–100 tons/h, (b) 1–60 tons, (c) 5–500 kg, (d) 1–300 tons
(3) Which melting stock is used for thermal insulation of molten metal?
       (a) Flux, (b) scrap, (c) slag-forming material, (d) ingots
(4) Which melting stock is used for degassing, cleaning, and grain refining?
       (a) Flux, (b) scrap, (c) slag-forming material, (d) ingots
(5) Which technique in the EAF process is the most beneficial for environment and economy?
       (a) Hot metal charging, (b) flux charging, (c) scrap charging, (d) coal charging
(6) Which type of alloy requires the addition of inoculant for microstructural control?
       (a) Steel, (b) cast iron, (c) aluminum alloy, (d) brass
(7) Which type of furnace involves the use of copper coils in its construction?
       (a) EAF, (c) crucible furnace, (c) cupola, (d) induction furnace

4.2. Explain why the electrodes of EAF are made of graphite?
4.3. Draw a labeled sketch of EAF.
4.4. (a) Why is induction furnace so named?
       (b) Explain the working of an induction furnace.
4.5. Explain the operation of a crucible furnace, with the aid of a labeled sketch.

   P-4.6. The data for the EAF circuit elements, for the peak power transferred from V_s to the load, are shown in Fig. 4.9. Determine the arc resistance for the peak power delivered to the load.

   P-4.7. By using the data in P-4.6, determine the peak power delivered to the arc in the EAF.

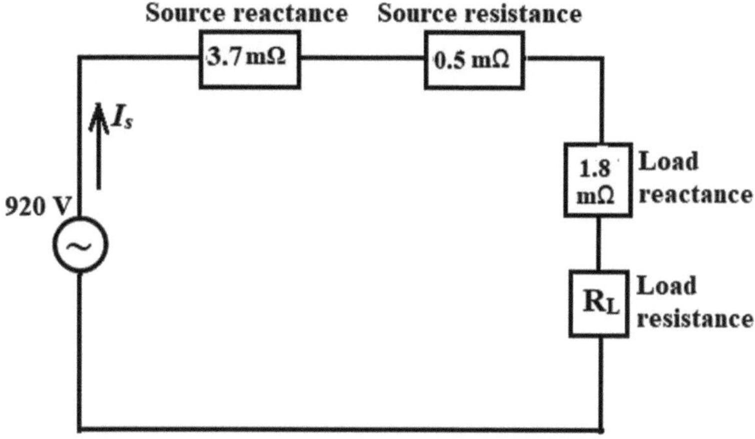

Fig. 4.9 The EAF circuit elements for P-4.6

P-4.8. A foundry melts 4 tons/day SG iron scrap in an induction furnace. Determine the monthly cost on electrical energy, if the unit price is $ 0.12/ kWh. Assume 1 month = 30 days.

P-4.9. The carbon emission of power per ton of molten steel with no addition of hot metal (HM) in an EAF is 140 kg. The electrical-energy consumption with use of hot metal ratio of 40% was determined to be 250 kWh/ton. Determine: (a) the carbon emission of the energy per ton of molten steel with HM ratio of 40%, and (b) the percent reduction in carbon emission due to the use of 40% HM ratio.

P-4.10. Spectrometry analysis of steel from a 12-ton ladle shows 0.20 wt.% Mn. Calculate the mass of Mn required to be added to achieve an aim composition of 1.3 wt.% Mn.

P-4.11. A 12-ton ladle steel contains 10.0% Cr at tap. Calculate the amount of ferrochrome required to be added to achieve a composition of 18.0% Cr. Ferrochrome contains 66.5% Cr. Typical recovery for Cr is 80%.

P-4.12. The rate of melting steel, in an induction furnace, is 550 kg/h. The furnace efficiency is 85%. Determine the gross power of the furnace. Hint: refer to the data in Table 4.2.

P-4.13. By using the data in P-4.12, determine the time required to melt 1 ton of aluminum in the same induction furnace.

P-4.14. The steel scrap, in a foundry, contains 97 wt.% Fe. Four (4) buckets are provided to charge the scrap to an EAF with a capacity of 82 tons/heat. What amount of scrap must be charged by use of a bucket?

P-4.15. An EAF supplies molten steel to a 2-strand CC plant that casts billets of dimensions (130 mm × 130 mm) in an hour at an average casting speed of 2200 mm/min. Determine the capacity of the furnace.

References

American Foundry Society. (2000). *Abrasion-resistant cast irons handbook*. AFS.

Huda, Z. (2020). *Metallurgy for physicists and engineers*. CRC Press.

Irawan, A., Kurniawan, T., Alwan, H., et al. (2022). An energy optimization study of the electric arc furnace from the steelmaking process with hot metal charging. *Heliyon, 8*(11), e11448.

Joseph, O., Peter, O., Irabodemeh, J. M., et al. (2016). Design and thermal analysis of crucible furnace for non–ferrous metal. *Journal of Information Engineering and Management, 6*(3), 1–9.

Kuhangana, T. C., Cheyns, K., Musambo, T. M., et al. (2024). Cottage industry as a source of high exposure to lead: A biomonitoring study among people involved in manufacturing cook-ware from scrap metal. *Environmental Research, 250*, 118493.

Liebermann, H. H. (Ed.). (1993). *Rapidly solidified alloys*. Dekker.

Martell-Chávez, F., Ramírez-Argáez, M., Llamas-Terres, A., & Micheloud-Vernackt, O. (2013). Theoretical estimation of peak arc power to increase energy efficiency in electric arc furnaces. *ISIJ Journal, 53*(5), 743–750.

Petrov, P. B. (2023). Material balance model for steel production in electric arc furnace—Melting period. *Journal Of Chemical Technology And Metallurgy, 58*(6), 1192–1198.

Yang, L.-z., Jiang, T., Li, G.-h., & Guo, Y.-f. (2016). Discussion of carbon emissions for charging hot metal in EAF steelmaking process. *High Temperature Materials and Processes, 36*(6), 615.

Chapter 5
Cold Blast Cupola

Nomenclature

H	Cupola's useful height
D	Inside diameter at the tuyeres' level
d_1	Inside diameter above the charging door
h_1	Height of shell from sill end to the top (excluding the hood)
h_d	Height of the charging door
h_2	Height of center-line of tuyeres from bottom plate
h	Effective depth of well
h_3	Height of tuyeres' level from slag hole
h_4	Height of the slag hole from the sand bottom
M	Metal holding capacity of the well
S	Fraction of the well's space to hold the metal
MR	Melting rate
Fe/C ratio	Metal/coke ratio
ρ_{CI}	Density of cast iron
m_{MC}	Mass of metal charge
m_i	Mass of element i in the constituent of the metal charge
Q_{ab}	Air-blast rate (m³/min)
$Q_{ab(o)}$	Air-blast rate (m³/m² • min)
m_C	Mass of carbon burnt (kg) per 100 kg of iron melted
V_{air}	Volume of air consumed (m³) per kg of carbon burnt
ΔH	Enthalpy change
ΔH_f	Latent eat of fusion
η_v	Degree of combustion
r_m	Initial modulus of metallic charge lumps

Z. Huda, *Metal Casting Engineering*, Mechanical Engineering Series,
https://doi.org/10.1007/978-3-031-84620-5_5

Abbreviations

ASHRAE	American Society of Heating Refrigeration and Air-Conditioning Engineers
BCIRA	British Cast Iron Research Association
CFM	Cubic feet per minute
CI	Cast iron
DoE	Department of Energy (Government of USA)
MC	Metal charge
PSC	Permanent split capacitor

5.1 The Cupola Furnace: Advantages and Limitations

The cupola is a melting furnace that is mainly used to produce molten cast iron. Although the cupola was primarily designed to convert pig iron to cast iron, its use has now been generalized to melt both pigs (solid pig iron) and cast-iron scrap. The cupola furnace has the following distinct advantages, which are responsible for its widespread use as a melting unit for cast iron: (a) the cupola is one of the few methods of melting that is continuous in its operation, (b) it has high melt rates, (c) the operating costs are relatively low, and (d) there is an ease of operation.

The main disadvantage of a cupola furnace is the pollution issuing from its chimney, which is costly to control. Another limitation is that the furnace is suitable for producing only large quantities of molten cast iron.

5.2 Cupola Furnace Construction

The cupola furnace is a tall cylindrical steel-shell that is lined with refractory bricks. It is fitted vertically with a charging door. The furnace is usually supported by four legs (see Fig. 5.1). The upper part of the cupola is provided with a charging door, through which metal, coke, and limestone (flux) are charged into the furnace. Cold air blast is admitted into the furnace through a series of tuyeres, which are provided around the periphery of the cupola. The top of the cupola furnace is equipped with a hood (or spark arrester), which arrests spark and burning particles and releases only gases into the environment. Cupola can be operated on different fuel-to-metal ratios, giving melt rates of approximately 1–30 tons per hour.

It is evident in Fig. 5.1 that the steel shell is lined with an inner edge of refractory brick and refractory-patch material. The diameter of the shell ranges from 45 to 200 cm, depending on the furnace's size. The cupola furnace is supported on four (4) cast iron legs that are mounted on a concrete base. There are two cast iron doors at the bottom of the furnace. Above the bottom, there is a sand bed; the molten iron

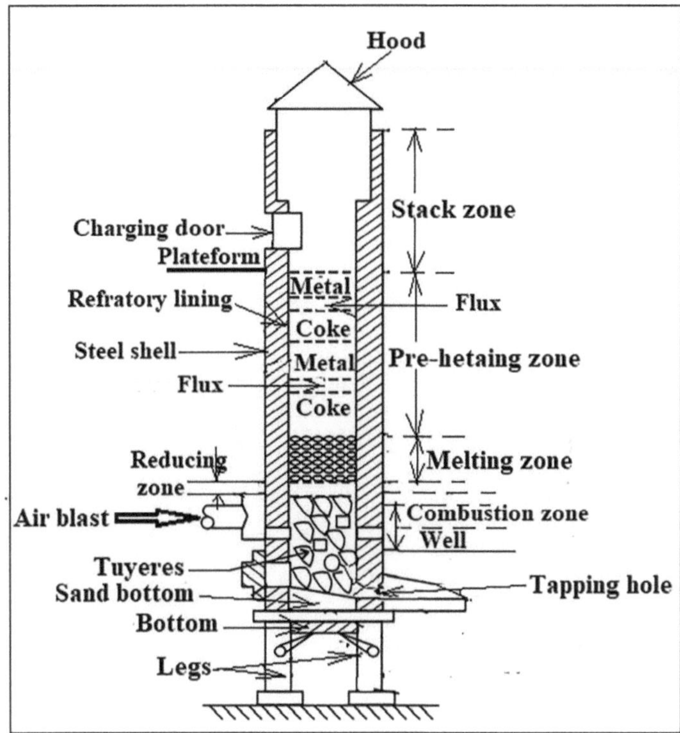

Fig. 5.1 Construction of a cupola furnace

flows on the sand bed, which is tapered (see Fig. 5.1). Above the sand bed, there is a series of tuyeres around the periphery of the cupola (see Fig. 5.2). Molten iron flows out though the *iron notch* or *tapping hole*. The slag is drained off through a *slag hole*, which is located 90–150 cm above the iron *notch*.

5.3   Working Principle of Cupola

The first step, in the operation of a cupola, is its clean up by opening the bottom drop doors to discharge the debris from previous use. Then the furnace is ignited by a small fire and coke is added gradually. This author enjoyed holding a scaffold of wood and getting the other end burnt with kerosene oil and lighter by his team. The burning end with flame was introduced into the well of the furnace. In general practice, coke is added to the level of 50 cm above the tuyeres. As soon as the coke bed is thoroughly ignited, alternate charges of metal charge, limestone, and coke are added in weighed proportions until level with the charging door (Ocheri, 2020). The metal charge consists of pig iron, cast-iron scrap, cast-iron returns (gates, risers, etc.), etc. Lime-stone is used as a flux at a proportion of 2–4% of the mass of metal

Fig. 5.2 A partial view of cupola (Courtesy: EC&S Inc., Birmingham, AL 35210, USA)

charge. The next step is to open tuyeres, and to supply a constant air volume to the combustion chamber. The carbon of the coke reacts with the oxygen of the air resulting in combustion (exothermic reaction), as follows:

$$C + O_2 \rightarrow CO_2, \Delta H = -410\,kJ\,/\,mol \tag{5.1}$$

The combustion of coke (exothermic reaction) (Eq. 5.1) releases a huge amount of heat resulting in a rise of temperature to about 1600–1800 °C. The high temperature causes the iron to melt and drain downwards. Then, CO_2 is partly reduced, consuming energy and coke, as follows:

$$CO_2 + 2C \rightarrow 2CO, \Delta H = +156\,kJ\,/\,mol \tag{5.2}$$

The endothermic reaction (Eq. 5.2) results in the drop of temperature to about 1200 °C in the reduction zone, which starts from the top of the combustion zone and extends to the top of the coke bed.

Finally, the metal charge transforms to a molten state in the melting zone; here, the temperature increases to 1600 °C to melt the metal charge. Then the molten metal drains down through the coke bed and reaches the well zone. Sufficient carbon content is picked by the molten metal in the melting zone. Limestone ($CaCO_3$) is decomposed into lime (CaO) and CO_2 gas. Lime reacts with other constituents to form slag, which is drained off through a slag hole. Cupola slag (CS) is used in cement concrete production. It has been reported that the concrete made with CS and recycled concrete aggregate (RCA) show satisfactory development and

Table 5.1 The inputs and outputs in a cupola operation

Inputs	Outputs
Pig iron	Molten cast iron
Cast iron scrap	Slag (liquid)
Cast iron returns (gates, risers, etc.)	Stack gases
Coke	
Flux/inoculants	
Air blast	

Table 5.2 Chemical composition of cast irons

Type of CI	% Carbon	% Silicon	% Manganese	% Sulfur	%Phosphorous
Gray CI	2.5–4.0	1.0–3.0	0.2–1.0	0.02–0.25	0.02–1.0
White CI	1.8–3.6	0.5–1.9	0.25–0.8	0.06–0.2	0.06–0.2

consistency in strength of concrete (Alibi & Mahachi, 2022). Recently, Lanjewar and co-researchers have reported the benefit of cupola slag (CS) as a partial replacement of coarse aggregate. The coarse aggregate was partially replaced by 30%, 35%, 40%, and 50% by CS without replacing the natural sand. The maximum compressive strength of 66.36 MPa for concrete was observed for 30% substitution on the 28th day. The results indicate that cupola slag can be used in concrete efficiently up to 30% and can provide a promising solution for reducing industrial waste landfill, ensuring a sustainable environment (Lanjewar et al., 2023).

The molten metal (cast iron) accumulates in the well and is tapped out via a tapping hole. The hot gases produced within the furnaces are escaped through the stack zone (see Fig. 5.1). To summarize, cupola operation transforms several inputs to outputs (see Table 5.1).

The main final product is molten cast iron. There are several types of cast iron (CI), including gray CI, white CI, ductile (or nodular) CI, compacted graphite CI, and the like (Huda, 2020). The range of compositions of gray CI and white CI are presented in Table 5.2 (Fisher, 2024).

5.4 Cupola's Design Analysis

5.4.1 Design Analysis for Furnace-Dimensions and Melting Rate

The cupola's design parameters include the inside diameter of the cupola at the tuyeres' level (D), the useful height (H), the production capacity (kg/h), and the like. The useful height H is the distance from the tuyeres' level to the charging door. The design parameters and components of cupola are shown in Fig. 5.3, and their nomenclatures are listed in Table 5.3.

Fig. 5.3 Schematic
showing cupola's design
parameters and
components

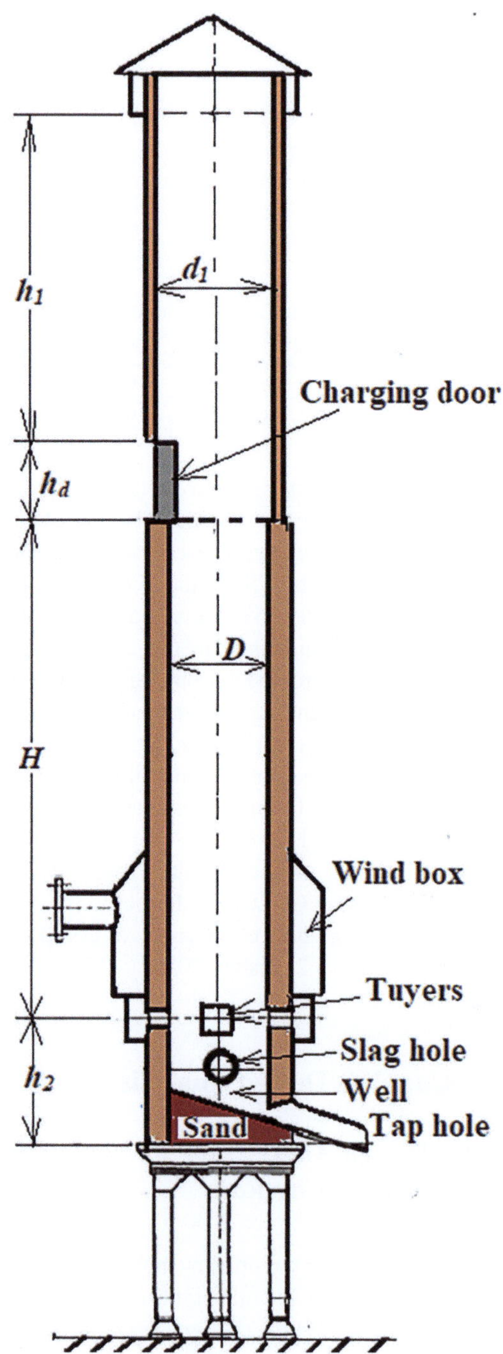

Table 5.3 Cupola's design parameters and their nomenclature

#	Design parameters	Symbol
1	Useful height	H
2	Inside diameter at the tuyere's level	D
3	Height of shell from sill end to the top (excluding the hood)	h_1
4	Size of the charging door	h_d
5	Inside diameter above the charging door	d_1
6	Height of center-line of tuyeres from bottom plate	h_2

Table 5.4 Melting rates (MR) per unit cross-sectional area for specified Fe/C ratios

Fe/C ratio	12	14	16	18
MR/area (tons/h/m²)	8.0	7.2	6.6	6.0

The inside diameter at the tuyere's level (D) determines the range of the melting rates of cupola, which in turn depends on the Fe/C (metal/coke) ratio. The Fe/C ratio, in general, ranges from 6 to 18. Table 5.4 presents the data for the average melting rate per unit cross-sectional area (MR/area) ranges for the corresponding Fe/C ratios (see Examples 5.1, 5.2, 5.3 and 5.4).

The cupola's H/D ratio depends on the strength of the coke to be charged, as follows:

$$H \cong 5.5\,D\left(\text{for high strength coke}\right) \tag{5.3a}$$

$$H \cong 3.75\,D\left(\text{for low strength coke}\right) \tag{5.3b}$$

The compressive strengths of low-strength coke lie in the range of 4–7 MPa, whereas those for high-strength coke can be as high as 23 MPa (see Example 5.5).

The portion of cupola between the tuyeres and the sand bottom is called the *well*. The effective depth of the well (h_2) is the height of the tuyeres' level from the sand bottom (see Fig. 5.2). The height of tuyeres' level (from slag hole), for D = 700 mm, is taken as h_3 = 250–280 mm. The height of the slag hole from the sand bottom, for D = 700 mm, is in the range of h_4 = 430–470 mm. By reference to Fig. 5.3, the height h_2 is:

$$h_2 = h_3 + h_4 \tag{5.4}$$

The significance of Eq. 5.4 is illustrated in Example 5.6.

The metal holding capacity (kg) of the well can be determined by multiplying the density of cast iron by the volume of the well that occupies the metal, as follows:

$$M = \rho_{CI} \cdot S \cdot \left(\frac{\pi}{4} \cdot D^2 \cdot h_2\right) \tag{5.5}$$

where M is the metal holding capacity of the well (mass of molten metal held in the well) (kg), ρ_{CI} is the density of liquid cast iron (kg/m^3), S is the fraction of the space to hold the metal (CI), D is the inside diameter of the cupola's well (m), and h_2 is the effective depth of the well (m) (see Example 5.7).

The height of the charging door, for D = 750 mm, is in the range of 350–400 mm (Indian Standard Institution, 1983). The cupola is extended 3–5 m above charging door, i.e. h_1 = 3–5 m (excluding the hood). Thus, the total height of cupola (excluding the legs and hood) can be calculated, with reference to Fig. 5.3, as follows:

$$\text{Total height} = h_1 + h_d + \text{H} + h_2 \tag{5.6}$$

The significance of Eq. 5.6 is illustrated in Example 5.8.

5.4.2 *Tuyeres' Design Analysis*

Tuyeres are the openings around the periphery of the cupola for introducing air blast to the coke bed inside the furnace. There are several tuyeres round the periphery of the cupola. The number of tuyeres in a series or the tuyere's ratio can be determined by (Nwajagu, 1994):

$$\text{Tuyere's ratio} = \frac{A}{A'} \tag{5.7}$$

where A is the inside cross-sectional area of the cupola at the tuyere's level, and A' is the total cross-sectional area of all tuyeres in a series. Tuyere's ratio values lie in the range of 4–15.

The cross-sectional area of one (each) tuyere, in a series, can be determined by:

$$\text{C.s area of one tuyere} = \frac{\text{Total cross sectional area of all tuyeres in a series}}{\text{Number of tuyeres in a series}}$$

$$= \frac{A'}{\text{Tuyeres' ratio}} \tag{5.8}$$

The significance of Eqs. 5.7 and 5.8 is illustrated in Example 5.9.

5.4.3 *Analysis for Air Blast: Volume Flow Rate*

Initially *air* blast *in cupola furnace* is mainly *supplied* by a blower, which is driven by 3 phase induction motor. The blast rate (volume flow rate of air blast) (in m^3/min) depends on a number of factors, including the (a) mass of coke burnt (kg) per 100 kg of iron melted, (b) volume of air consumed at STP (m^3) per kg of coke burnt, and (c) melting rate (tons/h). The mass of coke burnt (kg) per 100 kg of iron melted can be calculated by:

$$m_C = \left(\% \text{ age of coke in the charge} \times C \text{ content in the coke}\right) \\ - \% \text{ age of } C \text{ pick by the metal} \tag{5.9}$$

The blast flow rate can be determined by the following formula (BCIRA, 1979):

$$Q_{ab} = \frac{\left(m_C \bullet V_{air} \bullet MR\right)}{6} \tag{5.10}$$

where Q_{ab} is the air-blast flow rate (m³/min), m_C is the mass of coke burnt (kg) per 100 kg of iron melted, V_{air} is the volume of air consumed (m³) per kg of coke burnt, and MR is the melting rate (tons/h). The volume of air consumed (m³) per kg of coke burnt depends on the mass of carbon burnt (kg) per 100 kg of iron melted, as shown in the graphical plot in Fig. 5.4.

The significance of Eqs. 5.9 and 5.10 is illustrated in Examples 5.10 and 5.11.

The graphical plot in Fig. 5.4 indicates that V_{air} (the volume of air required to burn a unit mass of coke) initially varies inversely with m_C (kg of coke burnt per

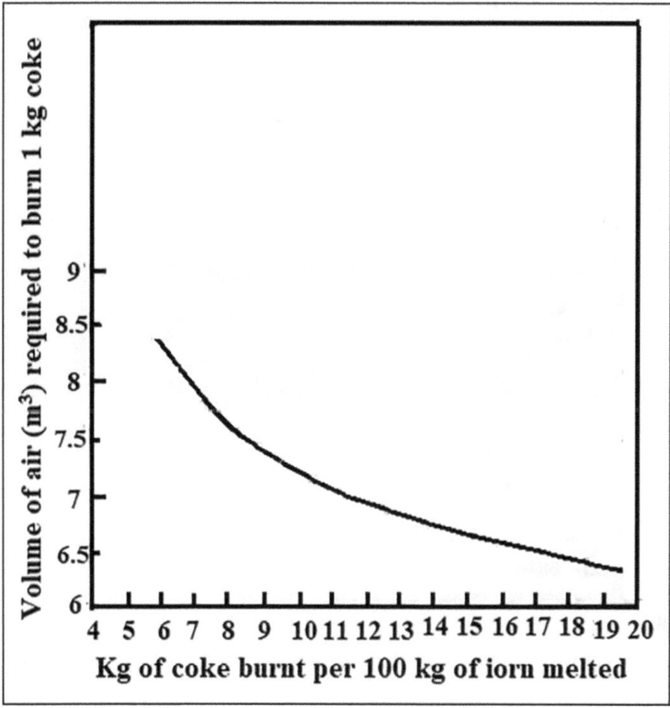

Fig. 5.4 The variation of V_{air} (m³) with m_C (kg)

100 kg of iron melted) with a 45° slope, and then the inverse variation occurs with a decrease in the slope of the curve.

5.4.4 Blower Fan Design Analysis

The electrical power consumption (and overall efficiency) of a blower motor depends on the speed at which the motor operates, the external static pressure difference across the blower, and the air-flow through the blower. The air-flow rate through a blower can be determined by (ASHRAE, 1997):

$$Q = \sqrt{\frac{P}{\alpha}} \tag{5.11}$$

where Q is the air-flow rate (ft^3/min or CFM), P is the external static pressure (inch water gage or in.w.g), and α is a constant (see *Example 5.12*). For furnaces with air handlers with permanent split capacitor (PSC) blower motors, the watts per CFM can be calculated by:

$$\frac{\text{Watts}}{\text{CFM}} = m_o + \left(m_1 * P\right) + \left(m_2 * P^2\right) \tag{5.12}$$

where m_o, m_1, and m_2 are the coefficients derived from second degree polynomial approximation (DOE, 2012) (see Example 5.13).

The power consumption of blower fan (P_{BF}) can be determined by:

$$P_{BF} = \left(\frac{\text{Watts}}{\text{CFM}}\right) * Q \tag{5.13}$$

where P_{BF} is the power consumption of blower fan (watts) (see Example 5.14).

5.5 Optimum Blast Flow Rate to Achieve the Maximum Metal Superheating

It was learnt in Chap. 2 that a successful casting requires good fluidity of the molten metal, which is achieved at high pouring temperatures (see Sect. 2.2.2). It means that extensively superheated molten cast iron from a cupola will produce good casting results. This section, therefore, focuses the analysis to achieve the maximum degree of superheating in the molten cast iron from a cupola.

The term "optimum blast volume flow rate" refers to the air-blast volume flow rate per unit internal cross-sectional area of the cupola [m^3/(m^2•s)] that will produce the maximum degree of molten cast iron overheating at standard operating conditions (coke consumption rate, blast air temperature, weight, shape and melting point of the lumps of metallic charge, height of the charge column above the level of tuyeres, etc.) (Khatemi & Longa, 2005).

The determination of the *optimum blast volume flow rate* involves the calculations for: (a) degree of combustion, η, (b) gas volume evolved from burning of 1 kg coke, V_g (m³/kg), (c) gas temperature at outlet from combustion zone, $T_{g,2}$ (°C), (d) decrease of gas temperature in the melting zone, $\Delta T_{g,2}$ (in K), (e) gas temperature at the outlet from combustion zone, $T_{g,3}$ (°C), etc.

The degree of combustion (η_v) can be determined by (Junghbluth, 1939):

$$\eta_v = \left(\frac{3.865}{K_w \bullet C_k} + 0.15 \right) \tag{5.14}$$

where K_w is the charge coke consumption (kg) per 100 kg of iron (in wt.%), and C_k is the mean specific heat of coke within the temperature range of $0–T_k$ (in kJ/kg K) (T_k is the coke temperature at the inlet to combustion zone, °C). The gas volume evolved from burning of 1 kg coke, V_g (m³/kg), can be determined by (Khatemi and Longa, 2005):

$$V_g = 0.054\left(100 + 0.65\eta_v\right) \cdot C_k \tag{5.15}$$

The significance of Eqs. 5.14 and 5.15 is illustrated in Example 5.15. The temperature of the gas at outlet from the combustion zone ($T_{g,2}$) (in °C) can be calculated by:

$$T_{g,2} = 1620 + 0.5 T_{ab} \tag{5.16}$$

where T_{ab} is the air-blast temperature. The decrease of gas temperature in the melting zone, $\Delta T_{g,2}$ (in K) can be determined by:

$$\Delta T_{g,2} = \frac{\Delta H_f}{V_g \bullet c_{g,2} \bullet K_w} \tag{5.17}$$

where ΔH_f is the latent heat of fusion of metal (J/kg), and $c_{g,2}$ is the mean specific heat of gas in the melting zone [in J/(m³·K)]. Once $T_{g,2}$ and $\Delta T_{g,2}$ have been determined, the temperature of the gas at outlet from the melting zone $T_{g,3}$ (in °C) can be determined by:

$$T_{g,3} = T_{g,2} - \Delta T_{g,2} \tag{5.18}$$

The significance of Eqs. 5.16, 5.17, and 5.18 is illustrated in Example 5.16.

The air volume consumed to burn 1 kg of coke at standard operating conditions, V_{air} (m³/kg) can be determined by:

$$V_{air} = 4.45 \times \left(1 + 0.01 \cdot \eta_v\right) \cdot C_k \qquad (5.19)$$

The significance of Eq. 5.19 is illustrated in Example 5.17.

The cast-iron scrap (pieces) is usually in the form of rectangular plates, prisms, cubes, and spheres. Let thickness, width, and length of a plate be a, b, and c, respectively. For cubes and spheres, $a = b = c$. The ratios of the dimensions can be expressed as:

$$m_b = \frac{b}{a} \qquad (5.20a)$$

$$m_c = \frac{c}{b} \qquad (5.20b)$$

The dimensionless coefficient for initial volume of metallic charge lump (ϕ_v) and the coefficient for initial surface area of the charge lump (ϕ_{sa}) can be calculated by (Longa & Freibe, 2001):

$$\varphi_v = \frac{1}{2} - \frac{1}{6 m_b} - \frac{1}{6 m_c} + \frac{1}{12 m_b \bullet m_c} \qquad (5.21a)$$

$$\varphi_{sa} = \frac{m_b \bullet m_c}{m_b + m_c + m_b \bullet m_c} \qquad (5.21b)$$

The coefficients' ratio for initial volume and surface area of the lumps (φ) can be calculated by:

$$\varphi = \frac{\varphi_v}{\varphi_{sa}} \qquad (5.22)$$

The significance of Eqs. 5.20a, 5.20b, 5.21a, 5.21b, and 5.22 is illustrated in Example 5.18.

The initial modulus of the metallic charge lumps (r_m) can be determined by:

$$r_m = \frac{V_{m,o}}{(SA)_{m,o}} \qquad (5.23)$$

where $V_{m,o}$ is the initial volume of the metallic charge lumps (m^3), and $(SA)_{m,o}$ is the initial surface area of the metallic charge lumps (m^2). The mean total modulus of metallic charge lumps in the melting zone ($\overline{r_m}$) can be determined by:

$$\overline{r_m} = r_m \bullet \varphi \qquad (5.24)$$

The significance of Eqs. 5.20a, 5.20b, 5.21a, 5.21b, 5.22, 5.23, and 5.24 is illustrated in Example 5.19.

The factors $K_{\rho,t}$ and K_ρ can be determined by (Longa, 1998; Longa & Freibe, 1999):

$$K_{\rho,t} = 1 + \left(\frac{K_w}{\varphi_v} \bullet \frac{\rho_{n,m}}{\rho_{n,k}} \right) \qquad (5.25a)$$

$$K_\rho = 1 + \left(K_w \bullet \frac{\rho_{n,m}}{\rho_{n,k}} \right) \qquad (5.25b)$$

where $\rho_{n,\,m}$ is the bulk density of metal, and $\rho_{n,\,k}$ is the bulk density of coke (see Example 5.20).

The specific heat of a mixture of materials filling the pre-heating zone (cast-iron scrap + coke +5 wt% limestone), $c_{m,\,3}$ (in J/(kg·K)), can be determined by:

$$c_{m,3} = 750 + 8K_w \qquad (5.26)$$

The significance of Eq. 5.26 is illustrated in Example 5.21.

Once the values of the variables mentioned in Eqs. 5.16, 5.17, 5.18, 5.19, 5.20a, 5.20b, 5.21a, 5.21b, 5.22, 5.23, 5.24, 5.25a, 5.25b and 5.26 have been calculated, the optimum blast volume flow rate $Q_{ab(o)}$ (in m³/(m²•s)) can be determined by (Khatemi & Longa, 2005):

$$Q_{ab(o)} = \frac{K_w \bullet V_{air} \bullet \rho_{n,m} \bullet \left(H - h_{cz} \right)}{\overline{r_m} \bullet \rho_m \left[\dfrac{\varphi \bullet K_{\rho,t} \bullet \Delta H_f}{\alpha_2 \bullet \Delta T_{g,2} \bullet} \bullet \ln\left(\dfrac{T_{g,2} - T_{m,f}}{T_{g,3} - T_{m,f}} \right) + \dfrac{K_\rho \bullet c_{m,3}}{\alpha_3 \bullet m_{3,1}} \bullet \ln\left(\dfrac{T_{g,4} - T_{m,o}}{T_{g,3} - T_{m,f}} \right) \right]} \qquad (5.27)$$

where the terms $m_{3,1}$ and $T_{g,4}$ can be calculated by the following formulas:

$$m_{3,1} = \frac{c_{m,3}}{V_g \cdot c_{g,3} \cdot K_w} \qquad (5.28)$$

$$T_{g,4} = T_{g,3} - m_{3,1} \bullet \left(T_{m,f} - T_{m,o} \right) \qquad (5.29)$$

where $c_{g,\,3}$ is the mean specific heat of gas in the pre-heating zone (J/(m³·K)), $T_{m,f}$ is the melting point of metallic charge (°C), $T_{m,\,o}$ is the initial temperature of the melting charge (°C), $T_{g,\,4}$ is the gas temperature on upper boundary of the pre-heating zone (°C), and the coefficients α_2 and α_3 are taken as 200 and 150 W/(m²·K), respectively. The significance of Eqs. 5.27, 5.28 and 5.29 is illustrated in Examples 5.22 and 5.23.

5.6   Cupola's Charge Calculations

5.6.1   Charging Materials in Cupola

The cupola's charging materials include metal charge (pig iron, cast-iron scrap, cast-iron returns, etc.), coke, and flux/inoculants. The flux normally constitutes 4% of the total charge weight. The weight of coke depends on the metal charge; normal metal/coke ratio varies from 10 to 16 (see Table 5.2). The weight of metal charge depends on the metal holding capacity of the well, which in turn relies on the inside diameter D (see Eq. 5.5).

5.6.2   Coke Bed Charge Calculation

It has been learnt in Sect. 5.3 that the first step in the operation of cupola is to discharge debris from the previous use by opening the bottom drop doors. If a cupola is examined after the discharge of the debris, a groove can be seen in the lining of the cupola. The coke bed charge should reach this groove's height above the sand bottom (see Fig. 5.5).

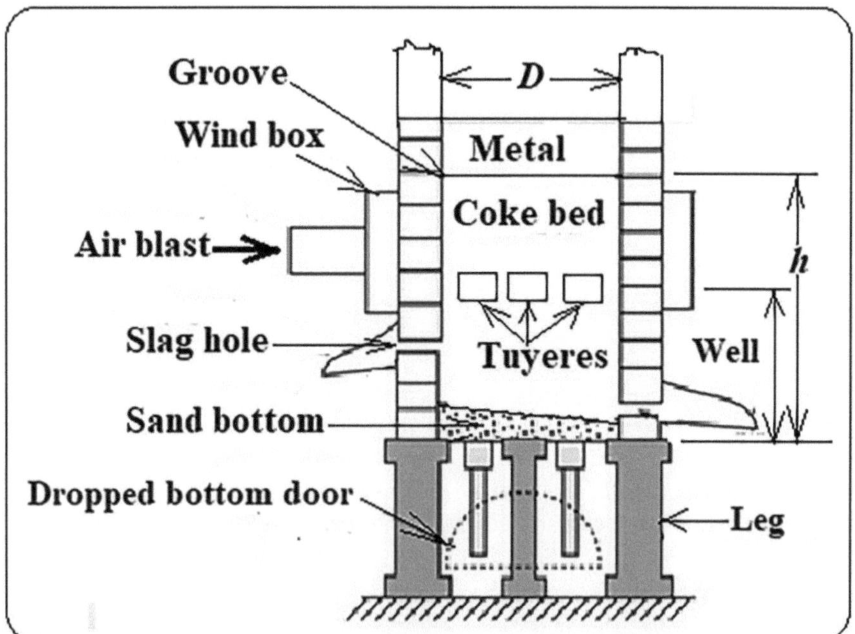

Fig. 5.5   A partial view of cupola showing the coke bed region

The coke-bed height h can be related to the inside diameter of the cupola by:

$$h = k \bullet D \tag{5.30}$$

where k lies in the range of 1.2–1.5. The mass of the coke bed charge can be determined, with reference to Fig. 5.5, as follows:

$$\text{Mass}_{\text{coke}} = \rho_{\text{coke}} \bullet \left(\frac{\pi}{4} \bullet D^2 \bullet h \right) \tag{5.31}$$

where $\text{Mass}_{\text{coke}}$ is the mass of coke bed (kg), ρ_{coke} is the density of coke (kg/m^3), D is the inside diameter of cupola at the tuyeres' level (m), and h is the height of the coke-bed from the sand bottom (m) (see Example 5.24).

5.6.3 Metal Charge Calculation

It has been learnt in Sect. 5.3 that the metal charge of a cupola furnace generally consists of pig iron, cast-iron scrap, and cast-iron returns (gates, riser, etc.). The quantity of metal charge depends on the metal holding capacity of the well (see Eq. 5.5).

The mass of an element i in a metallic constituent of the charge can be determined by the following formula (OTCA, 1984):

$$m_i = m_{\text{MC}} \cdot (\text{fraction of the constituent in the MC})$$
$$\cdot (\text{fraction of } i \text{ in the constituent}) \tag{5.32}$$

where m_i is the mass of element i in the constituent (kg), m_{MC} is the total mass of the metal charge (kg), and MC = metal charge. Since the metal charge consists of several constituents (pig iron, CI scrap, CI returns, etc.), the mass of element i in all metal constituents should be determined one-by-one by use of Eq. 5.32, and then these masses should be summed up to obtain the total mass of element i in the charge by using the following formula:

$$\text{Percentage of element } i \text{ in the final} \left(\text{molten} \right) \text{cast iron} = \frac{m_i}{m_{MC}} \times 100 \tag{5.33}$$

Then the percentage of the elemental pick up/loss should be added/subtracted to the value obtained by Eq. 5.33. This will give the content of element i in the final molten cast iron.

Since cast iron contains carbon (C), silicon (Si), manganese (Mn), sulfur (S), and phosphorous (P), the final content of each of these element may be determined one-by-one by using Eqs. 5.32 and 5.33; thus, the foundry engineer can predict the final analysis of the cupola product (see Examples 5.25, 5.26, 5.27, 5.28, and 5.29).

5.7 Worked Numerical Examples in Cold-Blast Cupola

Example 5.1 Designing a Cupola Furnace by Determining Its Inside Diameter
It is recommended to use a metal/coke ratio of 14 for a cupola furnace operation.
The required melting rate is 900 kg/h. Design the cupola by determining the inside
diameter at the tuyere's level. Hint: refer to Table 5.4.

Solution
By reference to the data in Table 5.4 for the Fe/C ratio of 14, $\dfrac{MR}{Area} = 8$ tons/h/m^2.

$$\text{Melting rate} = MR = 900\,kg\,/\,h = 0.90\,tons\,/\,h$$

$$\text{Cross sectional area} = \frac{MR}{8} = \frac{0.9}{8} = 0.1125\,m^2$$

$$\text{C.s.a} = \frac{\pi}{4} \cdot D^2 = 0.1125\,m^2$$

$$D^2 = 0.1125 \times \frac{4}{\pi} = 0.14324\,m^2$$

$$D = \sqrt{0.14324} = 0.3785\,m = 378.5\,mm$$

The inside diameter of the cupola at the tuyere's level = 378.5 mm.

Example 5.2 Determining the Outer Diameter of Cupola at the Tuyere's Level
By using the data in Example 5.1, determine the outer diameter at the tuyere's level
for the cupola. The refractory brick's thickness at the tuyere's level is 8 mm. The
shell is constructed by 5-mm-thick steel plates.

Solution
$t_{steel} = 5$ mm, $t_{ref} = 8$ mm, $D = 378.5$ mm, $D_o = ?$
   This problem can be solved with the aid of the sketch, as shown in Fig. 5.6.
   By reference to Fig. 5.6, the outer diameter (D_o) can be determined as follows:

$$D_o = D + 2 \cdot \left(t_{steel} + t_{ref}\right) = 378.5 + 2 \times \left(5 + 8\right) = 404.5\,mm$$

**Example 5.3 Determination of the Melting Rate of Iron and the Mass of
Molten Iron Tapped**
The inside diameter of a cupola at the tuyere's level is 700 mm. The metal/coke
ratio, to be used, is 16. Determine the: (a) melting rate of iron, and (b) mass of molten iron tapped for 2 taps in an hour.

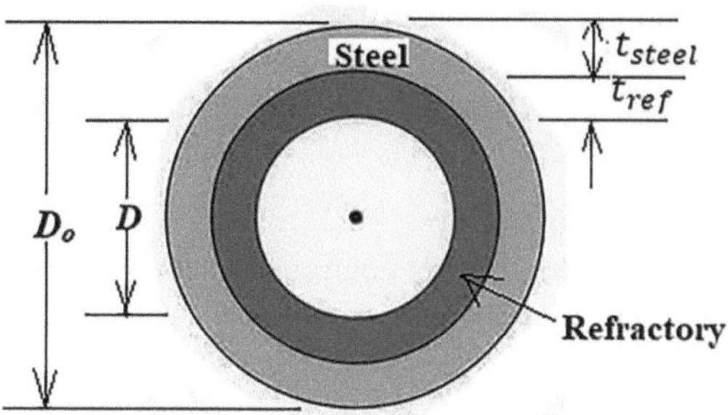

Fig. 5.6 The sketch (top view of cupola) for Example 5.2

Solution

D = 700 mm, Fe/C ratio = 16, MR =?

$$\text{Cross sectional area} = \frac{\pi}{4} \cdot D^2 = 0.785 \times 700^2 = 384650\,\text{mm}^2 \cong 0.38465\,\text{m}^2$$

(a) By reference to Table 5.4, MR/area = 6.6 tons/h/m².

$$\frac{MR}{0.38465} = 6.6$$

$$MR = 6.6 \times 0.38465 = 2.5387\,\text{ton}\,/\,\text{h}$$

(b) Mass of iron melted in 1 h = 2.5387 tons.

There are 2 taps per hour. The mass of molten iron tapped after 30 min $= \dfrac{2.53875}{2}$ = 1.269 tons.

Example 5.4 Determination of Coke Consumption Rate in a Cupola

By using the data in Example 5.3, determine the mass of coke consumed per hour.

Solution

MR = iron melted in an hour = 2.54 tons, Metal/coke ratio = 16.

$$\frac{\text{Mass of iron melted per hour}}{\text{Mass of coke consumed per hour}} = 16$$

$$\frac{\text{Mass of iron melted per hour}}{16} = \text{Mass of coke consumed per hour}$$

$$\text{Mass of coke consumed per hour} = \frac{2.54\,\text{tons}}{16} = 0.15875\,\text{ton} = 158.75\,\text{kg}$$

Example 5.5 Determination of the Useful Height of a Cupola Furnace
By using the data in Example 5.3, determine the useful height of the cupola, which is designed for use of coke with a compressive strength of 20 MPa.

Solution
D = 700 mm, H =?
  Since the 20-MPa compressive strength lies in the range of a high-strength coke, Eq. 5.3a is applicable, as follows:

$$H \cong 5.5\,D = 5.5 \times 700 = 3850\,mm = 3.85\,m$$

The useful height of the cupola furnace = H = 3.85 m.

Example 5.6 Determining the Effective Depth of the Well of a Cupola
The inside diameter of a cupola at the tuyere's level is 700 mm. The height of tuyeres' level (from slag hole), for a cupola furnace, is 260 mm. The height of the slag hole from the sand bottom is 450 mm. Determine the effective depth of the well.

Solution
h_3 = 260 mm, h_4 = 450 mm, h_2 =?
  By using Eq. 5.4,

$$h_2 = h_3 + h_4 = 260 + 450 = 710\,mm = 0.71\,m$$

The effective depth of the well = h_2 = 0.71 m.

Example 5.7 Determining the Metal Holding Capacity of a Cupola's Well
By using the data in Example 5.6, determine the metal holding capacity of the well of the cupola. The fraction of the well's space to hold the metal is 40%.

Solution
D = 700 mm = 0.70 m, h_2 = 0.71 m, S = 40% = 0.40, ρ_{liq} of CI = 7015 kg/m³ (see Table 3.1).
  By using Eq. 5.5,

$$M = \rho_{CI} \cdot S \cdot \left(\frac{\pi}{4} \cdot D^2 \cdot h_2 \right) = 7015 \times 0.40 \times \left(\frac{\pi}{4} \times 0.70^2 \times 0.71 \right) = 2806 \times 0.2731$$

$$M = 766.32\,kg$$

The metal holding capacity of the well of the cupola = 766.32 kg.

Example 5.8 Determining the Total Height of Cupola Furnace
By using the design data in Examples 5.5 and 5.6, determine the total height of the cupola. The height from the charging-door sill to the cupola top (excluding hood) is 3 m. Take the average of the other heights.

Solution
h_1= 3 m, h_2 = 0.71 m, h_d = 375 m = 0.375 m, H = 3.85 m.

By using Eq. 5.6,

$$\text{Total height of cupola} = h_1 + h_d + \text{H} + h_2 = 3 + 0.375 + 3.85 + 0.71 = 7.93\,\text{m}$$

Example 5.9 Tuyere's Design for a Cupola Furnace
By using the design data in Examples 5.3, 5.4, 5.5, 5.6, 5.7, and 5.8, determine the diameter of each tuyere if the tuyeres' ratio is 8.

Solution
A = 0.385 m^2, tuyeres' ratio = 8.
   By using Eqs. 5.7 and 5.8,

$$\text{Tuyere's ratio} = \frac{A}{A'} = 8$$

$$A' = \frac{A}{8} = \frac{0.385}{8} = 0.048125\,\text{m}^2$$

$$\text{C.s area of one tuyere} = \frac{\text{Total cross sectional area of all tuyeres in a series}}{\text{Number of tuyeres in a series}} = \frac{A'}{8}$$

$$= \frac{0.048125}{8} = 0.006\,\text{m}^2$$

$$\frac{\pi}{4} \cdot d_t^2 = 0.006$$

$$d_t^2 = 0.006 \times \frac{4}{\pi} = 0.0076394$$

$$d_t = \sqrt{0.0076394} = 0.087\,\text{m} = 87\,\text{mm}$$

The diameter of each tuyere = 87 mm.

Example 5.10 Determining the Mass of Carbon Burnt (kg) per 100 kg of Iron Melted
The percentage of coke, in a cupola-furnace charge, is 16. The carbon (C) content of the coke is 88%. The percentage of carbon pick by the metal is 0.45. Determine the mass of carbon burnt (kg) per 100 kg of iron melted.

Solution
By using Eq. 5.9,

$$m_C = \left(\% \text{ age of coke in the charge} \times C \text{ content in the coke}\right)$$
$$- \% \text{ age of } C \text{ pick by the metal}$$

$$m_C = \left(16 \times 0.88\right) - 0.45 = 13.63 \, \text{kg}$$

The mass of carbon burnt per 100 kg of iron melted $= m_C = 13.63$ kg.

Example 5.11 Determining the Air Blast Rate for the Cupola Operation
By using the data in Example 5.10, determine the blast rate for the cupola operation.
The melt rate is 10 tons/h. Hint: refer to Fig. 5.4.

Solution
$m_C = 13.63$ kg, MR = 10 tons/h, blast rate = ?
  By reference to Fig. 5.4 for the mass of carbon $m_C = 13.63$ kg, the corresponding
volume of air (V_{air}) is 6.7 m^3.
  By using Eq. 5.10,

$$Q_{ab} = \frac{\left(m_C \cdot V_{air} \cdot MR\right)}{6} = \frac{\left(13.63 \times 6.7 \times 10\right)}{6} = 152.2 \ \text{m}^3 / \text{min}$$

The blast rate = 152.2 m^3/min.

**Example 5.12 Determining the Air Flow Rate Through the Blower of
a Cupola**
The external static pressure for a high-fire cupola is 0.5 in.w.g. Determine the air-
flow rate through the blower of a high-fire cupola furnace. Take the constant
$\alpha = 4.0 \times 10^{-7}$.

Solution
P = 0.5 in.w.g, $\alpha = 4.0 \times 10^{-7}$, Q = ?
  By using Eq. 5.11,

$$Q = \sqrt{\frac{P}{\alpha}} = \sqrt{\frac{0.5}{4.0 \times 10^{-7}}} = \sqrt{\frac{5 \times 10^6}{4.0}} = 1118 \, \text{ft}^3 / \text{min}$$

The air-flow rate through the blower = 1118 CFM.

Example 5.13 Determining the Watts per CFM for the Blower Fan
The external static pressure for a high-fire cupola is 0.5 in.w.g. The coefficients
derived from second degree polynomial approximation for the 2-tons furnace are
0.395, −0.16, and 0.258, respectively. Determine the $\dfrac{\text{Watts}}{\text{CFM}}$ for the blower fan.

Solution
P = 0.5 in.w.g, $m_o = 0.395$, $m_1 = -0.16$, $m_2 = 0.258$.

$$\frac{\text{Watts}}{\text{CFM}} = m_o + (m_1 * \text{P}) + (m_2 * P^2) = 0.395 + (-0.16 \times 0.5) + (0.258 \times 0.5^2)$$

$$\frac{\text{Watts}}{\text{CFM}} = 0.395 - 0.08 + 0.0645 = 0.38 \ \text{watts/cfm}$$

Example 5.14 Determining the Power Consumption of the Blower Fan

By using the data in Examples 5.12 and 5.13, determine the power consumption of the blower fan.

Solution

$\dfrac{\text{Watts}}{\text{CFM}} = 0.38$ watts/cfm, $Q = 1118$ cfm, $P_{BF} = ?$

By using Eq. 5.13,

$$P_{BF} = \left(\frac{\text{Watts}}{\text{CFM}}\right) * Q = 0.38 \times 1118 = 425 \ \text{watts}$$

The power consumption of the blower fan = 425 watts.

Example 5.15 Determining the Volume of the Gas Evolved from Burning of 1 kg Coke

A cold-blast cupola is charged in a single-row coke. The charge coke consumption is 13 kg coke per 100 kg iron, i.e., 13 wt.%. The mean specific heat of coke within the temperature range of 0–1620 °C is 0.860 kJ/(kg•K). Determine the volume of the gas evolved from burning of 1 kg coke.

Solution

$K_w = 13$ wt.% $= 0.13$, $C_k = 0.86$ kJ/(kg•K), $V_g = ?$

By using Eqs. 5.14 and 5.15,

$$\eta_v = \left(\frac{3.865}{K_w \cdot C_k} + 0.15\right) = \left(\frac{3.865}{0.13 \times 0.86} + 0.15\right) = 34.72$$

$$V_g = 0.054(100 + 0.65\eta_v) \cdot C_k = 0.054 \times (100 + 0.65 \times 34.72) \times 0.86 = 5.7 \, \text{m}^3 / \text{kg}$$

The volume of the gas evolved from burning of 1 kg coke = 5.7 m³

Example 5.16 Determining the Gas Temperature at Outlet from the Melting Zone

By using the data in Example 5.15, determine the gas temperature at outlet from the melting zone. The cold-blast cupola is being operated at an air-blast temperature of 0 °C. The mean specific heat of gas in the melting zone (in the temperature range of 1400–1600 °C) is 1708 J/(m³·K). The latent heat of fusion of iron is 272,000 J/kg.

Solution

$T_{ab} = 0$ °C, $K_w = 0.13$, $V_g = 5.7$ m³/kg, $\Delta H_f = 272{,}000$ J/kg, $c_{g,2} = 1708$ J/(m³·K)

By using Eq. 5.16,

$$T_{g,2} = 1620 + 0.5\,T_{ab} = 1620 + 0.5 \times 0 = 1620°\mathrm{C} = 1893\,\mathrm{K}$$

By using Eqs. 5.17 and 5.18,

$$\Delta T_{g,2} = \frac{\Delta H_f}{V_g \cdot c_{g,2} \cdot K_w} = \frac{272000}{5.7 \times 1708 \times 0.13} = 215\,\mathrm{K}$$

$$T_{g,3} = T_{g,2} - \Delta T_{g,2} = 1893 - 215 = 1678\,\mathrm{K} = 1405°\mathrm{C}$$

The gas temperature at the outlet from the melting zone $= T_{g,3} = 1405$ °C.

Example 5.17 Determining the Volume of Air to Burn 1 kg Coke for Maximum Superheat

By using the data in Example 5.15, determine the volume of the air consumed to burn 1 kg of coke at standard operating conditions.

Solution

$\eta_v = 34.72$, $C_k = 0.86$ kJ/(kg•K), $V_{air} = ?$

$$\begin{aligned}
V_{air} &= 4.45 \times \left(1 + 0.01 \cdot \eta_v\right) \cdot C_k \\
&= 4.45 \times \left(1 + 0.01 \times 34.72\right) \times 0.86 \\
&= 5.15\,\mathrm{m}^3 / \mathrm{kg}
\end{aligned}$$

Example 5.18 Determining the Coefficients' Ratio for the Initial Volume and Surface Area

A cast-iron scrap-plate has dimensions $a = 0.05$ m, $b = 0.2$ m, and $c = 0.3$ m. Determine the coefficients' ratio for the initial volume and surface area of the metallic charge lump.

Solution

By using Eqs. 5.20a, 5.20b, 5.21a, 5.21b, and 5.22,

$$m_b = \frac{b}{a} = \frac{0.2}{0.05} = 4 \qquad\qquad m_c = \frac{c}{b} = \frac{0.3}{0.2} = 1.5$$

$$\varphi_v = \frac{1}{2} - \frac{1}{6m_b} - \frac{1}{6m_c} + \frac{1}{12 m_b \cdot m_c} = \frac{1}{2} - \frac{1}{6\times4} - \frac{1}{6\times1.5} + \frac{1}{12\times4\times1.5} = 0.361$$

$$\varphi_{sa} = \frac{m_b \cdot m_c}{m_b + m_c + m_b \cdot m_c} = \frac{4\times1.5}{4+1.5+4\times1.5} = \frac{6}{4+1.5+6} = 0.522$$

$$\varphi = \frac{\varphi_v}{\varphi_{sa}} = \frac{0.361}{0.522} = 0.691$$

Example 5.19 Determining the Mean Total Modulus of Metallic Charge Lumps

By using the data in Example 5.18, determine the mean total modulus of metallic charge lumps in the melting zone. The initial volume of the metallic charge lumps, in a cupola, is 0.003 m³. The initial surface area of the metallic charge lump is 0.17 m².

Solution
$V_{m,o} = 0.003$ m³, $(SA)_{m,o} = 0.17$ m², $\varphi = 0.691$
   By using Eqs. 5.23 and 5.24,

$$r_m = \frac{V_{m,o}}{(SA)_{m,o}} = \frac{0.003}{0.17} = 0.0176 \text{ m}$$

$$\overline{r_m} = r_m \cdot \varphi = 0.0176 \times 0.691 = 0.0121$$

Example 5.20 Determining the Factors $K_{\rho,t}$ and K_ρ for Cold-Blast Cupola

By using the data in Examples 5.15, 5.16, 5.17, and 5.18, determine the factors $K_{\rho,t}$ and K_ρ for cast-iron scrap remelting in the cold-blast cupola. The bulk densities of the metal and coke are 2500 kg/m³ and 500 kg/m³, respectively.

Solution
$K_w = 13$ wt.% $= 0.13$, $\varphi_v = 0.361$, $\rho_{n,m} = 2500$ kg/m³, $\rho_{n,k} = 500$ kg/m³
   By using Eqs. 5.25a and 5.25b),

$$K_{\rho,t} = 1 + \left(\frac{K_w}{\varphi_v} \cdot \frac{\rho_{n,m}}{\rho_{n,k}} \right) = 1 + \left(\frac{0.13}{0.361} \times \frac{2500}{500} \right) = 2.8$$

$$K_\rho = 1 + \left(K_w \cdot \frac{\rho_{n,m}}{\rho_{n,k}} \right) = 1 + \left(0.13 \times \frac{2500}{500} \right) = 1.65$$

Example 5.21 Determine the Specific Heat of a Mixture of Materials in the Pre-heating Zone

By using the data in Example 5.20, determine the specific heat of a mixture of materials filling the pre-heating zone (cast-iron scrap + coke + $CaCO_3$ making 5 wt.% of the coke charge).

Solution
$K_w = 13$ wt.% $= 0.13$, $c_{m,3} = ?$

By using Eq. 5.26,

$$c_{m,3} = 750 + 8K_w = 750 + 8 \times 0.13 = 751\,\mathrm{J}/(\mathrm{kg}\cdot\mathrm{K})$$

Example 5.22 Determining the Gas Temperature on Upper Boundary of Pre-heating Zone

By using the data in Examples 5.15, 5.16, 5.17, 5.18, 5.19, 5.20, and 5.21, determine the (a) term $m_{3,1}$, and (b) gas temperature on upper boundary of the pre-heating zone. The melting point of the metallic charge is 1150 °C. The initial temperature of metallic charge is 20 °C. The mean specific heat of gas in the pre-heating zone is 1600 J/(m³·K).

Solution

$c_{m,3} = 751$ J/(kg·K), $V_g = 5.7$ m³/kg, $c_{g,3} = 1600$ J/(m³·K), $K_w = 0.13$, $T_{m,f} = 1150$ °C, $T_{m,o} = 20$ °C, $T_{g,3} = 1405$ °C, $m_{3,1} = ?$, $T_{g,4} = ?$

By using Eqs. 5.28 and 5.29,

$$m_{3,1} = \frac{c_{m,3}}{V_g \cdot c_{g,3} \cdot K_w} = \frac{751}{5.7 \times 1600 \times 0.13} = 0.633$$

$$T_{g,4} = T_{g,3} - m_{3,1} \cdot (T_{m,f} - T_{m,o}) = 1405 - 0.633 \times (1150 - 20) = 989^\circ\mathrm{C}$$

Example 5.23 Determining the Optimum Blast Volume Flow Rate for Cupola

By using the data in Examples 5.15, 5.16, 5.17, 5.18, 5.19, 5.20, 5.21, and 5.22, determine the optimum blast volume flow rate (in m³/(m²•s)). The useful height of the cupola is 5 m, and the height of the combustion zone is 0.3 m. The density of metallic charge lumps is 7000 kg/m³.

Solution

$K_w = 0.13$, $V_{air} = 5.15$ m³/kg, $\rho_{n,m} = 2500$ kg/m³, $H = 5$ m, $h_{cz} = 0.3$ m, $\overline{r}_m = 0.0121$, $\rho_m = 7000$ kg/m³, $\varphi = 0.691$, $K_{\rho,t} = 2.8$, $K_\rho = 1.65$, $\Delta H_f = 272{,}000$ J/kg, $\alpha_2 = 200$ W/(m²·K), $\alpha_3 = 150$ W/(m²·K), $\Delta T_{g,2} = 215$ K, $T_{g,2} = 1620$ °C, $T_{g,3} = 1405$ °C, $T_{m,f} = 1150$ °C, $T_{m,o} = 20$ °C, $c_{m,3} = 751$ J/(kg·K), $m_{3,1} = 0.633$, $T_{g,4} = 989$ °C.

By using Eq. 5.27,

$$Q_{ab(o)} = \frac{K_w \bullet V_{air} \bullet \rho_{n,m} \bullet (H - h_{cz})}{\overline{r}_m \bullet \rho_m \bullet \left[\dfrac{\varphi \bullet K_{\rho,t} \bullet \Delta H_f}{\alpha_2 \bullet \Delta T_{g,2}} \bullet \ln\left(\dfrac{T_{g,2} - T_{m,f}}{T_{g,3} - T_{m,f}}\right) + \dfrac{K_\rho \bullet c_{m,3}}{\alpha_3 \bullet m_{3,1}} \bullet \ln\left(\dfrac{T_{g,4} - T_{m,o}}{T_{g,3} - T_{m,f}}\right) \right]}$$

$$= \frac{0.13 \times 5.15 \times 2500 \times (5 - 0.3)}{0.0121 \times 7000 \times \left[\dfrac{0.691 \times 2.8 \times 272000}{200 \times 215} \times \ln\left(\dfrac{1620 - 1150}{1405 - 1150}\right) + \dfrac{1.65 \times 751}{150 \times 0.633} \times \ln\left(\dfrac{989 - 20}{1405 - 1150}\right) \right]}$$

$$Q_{ab(o)} = \frac{7866.625}{84.7 \times \left[12.239 \times \ln 1.843 \ + \ 13.050 \times \ln 3.8\right]}$$

$$= \frac{7866.625}{84.7 \times \left(7.4828 + 17.4217\right)} = \frac{7866.625}{2109.416} = 3.73$$

The optimum air-blast volume flow rate = 3.73 m³/(m²•s) = 223.8 m³/(m²•min).

Example 5.24 Determining the Mass of the Coke-Bed Charge
The inside diameter of cupola at the tuyeres' level is 700 mm. The density of coke is 480 kg/m³. Determine the mass of the coke-bed charge.

Solution
D = 700 mm = 0.7 m, ρ_{coke} = 480 kg/m³
  By using Eqs. 5.30 and 5.31,

$$h = k\,D = 1.3 \times 0.7 = 0.91\,m$$

$$Mass_{coke} = \rho_{coke} \cdot \left(\frac{\pi}{4} \cdot D^2 \cdot h\right) = 480 \times 0.785 \times 0.7^2 \times 0.91 = 168\,kg$$

The mass of the coke-bed charge = 168 kg

Example 5.25 Determination (Prediction) of Carbon Content in the Final Molten Cast Iron
A cupola's metal charge has a mass of 1 ton. The metal charge is made up of the following constituents: pig iron (35%), cast-iron scrap (30%), and cast-iron returns (35%). The contents of impurity elements (carbon, silicon, manganese, sulfur, phosphorous) in each constituent are presented in Table 5.5. Determine the carbon content in the final cupola product (molten cast iron). The carbon pick up from coke is 0.14%.

Solution
m_{MC} = 1 ton = 1000 kg. The mass of carbon in each constituent can be calculated, as follows.

For Carbon in Pig Iron
  Fraction of pig iron in the metal charge = 35% = 0.35

Table 5.5 Impurity element contents in each metal-charge constituents

Constituent	Pig iron	Cast-iron scrap	Cast-iron returns
Carbon (C)	3.6%	3.5%	3.4%
Silicon (Si)	2.7%	2.2%	2.4%
Manganese (Mn)	0.68%	0.5%	0.6%
Sulfur (S)	0.017%	0.032%	0.034%
Phosphorous (P)	0.14%	0.19%	0.15%

By using Eq. 5.32,

$$m_{C-1} = m_{MC} \cdot (\text{fraction of pig iron in the MC}) * (\text{fraction of carbon in the pig iron})$$
$$m_{C-1} = 1000 \times 0.35 \times 0.036 = 12.6 \text{ kg}$$

For Carbon in Cast Iron Scrap
Fraction of cast-iron scrap in the metal charge = 30% = 0.30

$$m_{C-2} = m_{MC} \cdot (\text{fraction of cast iron scrap in the MC})$$
$$* (\text{fraction of carbon in the cast iron scrap})$$
$$m_{C-2} = 1000 \times 0.30 \times 0.035 = 10.5 \text{ kg}$$

For Carbon in Cast Iron Returns
Fraction of cast-iron returns in the metal charge = 35% = 0.35

$$m_{C-3} = m_{MC} \cdot (\text{fraction of cast iron returns in the MC})$$
$$* (\text{fraction of carbon in the CI returns})$$
$$m_{C-3} = 1000 \times 0.35 \times 0.034 = 11.9 \text{ kg}$$

For Carbon in Total Metal Charge

$$\text{Mass of carbon in the metal charge} = m_{\text{carbon}} = m_{C-1} + m_{C-2} + m_{C-3}$$
$$m_{\text{carbon}} = 12.6 + 10.5 + 11.9 = 35 \text{ kg}$$

By using Eq. 5.33,

$$\text{Percentage of carbon in the final} (\text{molten}) \text{cast iron} = \frac{m_{\text{carbon}}}{m_{MC}} \times 100$$
$$= \frac{35}{1000} \times 100 = 3.5$$

Actual percentage of carbon in the final (molten) cast iron = 3.5 + 0.14 = 3.64

Example 5.26 Determination (Prediction) of Silicon Content in the Final Molten Cast Iron

By using the data in Example 5.25, determine the silicon (Si) content in the final cupola product (molten cast iron). The silicon loss during melting is 12% of total silicon in the metal charge.

Solution
m_{MC} = 1 ton = 1000 kg. The mass of silicon in each constituent can be calculated, as follows.

For Silicon in Pig Iron

$$m_{Si-1} = m_{MC} \cdot (\text{fraction of pig iron in the MC}) \cdot (\text{fraction of silicon in the pig iron})$$
$$m_{Si-1} = 1000 \times 0.35 \times 0.027 = 9.45 \text{ kg}$$

For Silicon in Cast Iron Scrap

$$m_{Si-2} = m_{MC} \cdot (\text{fraction of cast iron scrap in the MC})$$
$$\cdot (\text{fraction of silicon in the cast iron scrap})$$
$$m_{Si-2} = 1000 \times 0.30 \times 0.022 = 6.6 \text{ kg}$$

For Silicon in Cast Iron Returns

$$m_{Si-3} = m_{MC} \cdot (\text{fraction of cast iron returns in the MC})$$
$$\cdot (\text{fraction of silicon in the CI returns})$$
$$m_{Si-3} = 1000 \times 0.35 \times 0.024 = 8.4 \text{ kg}$$

For Silicon in Total Metal Charge

$$\text{Mass of silicon in the metal charge} = m_{\text{silicon}} = m_{Si-1} + m_{Si-2} + m_{Si-3}$$
$$m_{\text{silicon}} = 9.45 + 6.6 + 8.4 = 24.45 \text{ kg}$$

$$\text{Percentage of silicon in the final (molten) cast iron} = \frac{m_{\text{Silicon}}}{m_{\text{mc}}} \times 100$$

$$= \frac{24.45}{1000} \times 100 = 2.445$$

Actual percentage of silicon in the final (molten) cast iron = 2.445—(12% × 2.445) = 2.15

Example 5.27 Determination (Prediction) of Manganese Content in the Final Molten Metal

By using the data in Example 5.25, determine the manganese (Mn) content in the final cupola product (molten cast iron). The Mn loss during melting is 18% of total Mn in the metal charge.

Solution

For Manganese in Pig Iron

$$m_{Mn-1} = m_{MC} \cdot (\text{fraction of pig iron in the MC}) \cdot (\text{fraction of Mn in the pig iron})$$
$$m_{Mn-1} = 1000 \times 0.35 \times 0.0068 = 2.38 \text{ kg}$$

For Manganese in Cast Iron Scrap

$$m_{Mn-2} = m_{MC} \cdot \left(\text{fraction of cast\ iron scrap in the MC}\right)$$
$$\cdot \left(\text{fraction of Mn in the cast\ iron scrap}\right)$$
$$m_{Mn-2} = 1000 \times 0.30 \times 0.005 = 1.5 \ \text{kg}$$

For Manganese in Cast Iron Returns

$$m_{Mn-3} = m_{MC} \cdot \left(\text{fraction of cast\ iron returns in the MC}\right)$$
$$\cdot \left(\text{fraction of Mn in the CI returns}\right)$$
$$m_{Mn-3} = 1000 \times 0.35 \times 0.006 = 2.1 \ \text{kg}$$

For Manganese in Total Metal Charge

$$\text{Mass of manganese in the metal charge} = m_{Mn} = m_{Mn-1} + m_{Mn-2} + m_{Mn-3}$$
$$m_{Mn} = 2.38 + 1.5 + 2.1 = 5.98 \ \text{kg}$$

$$\text{Percentage of manganese in the final}\left(\text{molten}\right)\text{cast iron} = \frac{m_{Mn}}{m_{MC}} \times 100$$

$$= \frac{5.98}{1000} \times 100 = 0.598$$

Actual percentage of manganese in the final (molten) cast iron = 0.598—(18% × 0.598) = 0.49

Example 5.28 Determination (Prediction) of Sulfur Content in the Final Molten Metal

By using the data in Example 5.25, determine the sulfur (S) content in the final cupola product (molten cast iron). Assume 5% gain in sulfur content.

Solution

For Sulfur in Pig Iron

$$m_{S-1} = m_{MC} \cdot \left(\text{fraction of pig iron in the MC}\right) \cdot \left(\text{fraction of S in the pig iron}\right)$$
$$m_{S-1} = 1000 \times 0.35 \times 0.00017 = 0.059 \ \text{kg}$$

For Sulfur in Cast Iron Scrap

$$m_{S-2} = m_{MC} \cdot \left(\text{fraction of cast\ iron scrap in the MC}\right)$$
$$\cdot \left(\text{fraction of S in the cast\ iron scrap}\right)$$
$$m_{S-2} = 1000 \times 0.30 \times 0.00032 = 0.096 \ \text{kg}$$

For Sulfur in Cast Iron Returns

$$m_{S-3} = m_{MC} \cdot (\text{fraction of cast iron returns in the MC})$$
$$\cdot (\text{fraction of S in the CI returns})$$
$$m_{S-3} = 1000 \times 0.35 \times 0.00034 = 0.119 \text{ kg}$$

For Sulfur in Total Metal Charge

$$\text{Mass of sulfur in the metal charge} = m_S = m_{S-1} + m_{S-2} + m_{S-3}$$
$$m_S = 0.059 + 0.096 + 0.119 = 0.274 \text{ kg}$$

$$\text{Percentage of sulfur in the final (molten) cast iron} = \frac{m_S}{m_{MC}} \times 100$$

$$= \frac{0.274}{1000} \times 100 = 0.0274$$

Actual percentage of sulfur in the final (molten) cast iron $= 0.0274 + 0.05 = 0.077$

Example 5.29 Determination (Prediction) of Complete Analysis in the Final Molten Metal

By using the data in Examples 5.25, determine the phosphorous (P) content in the final cupola product (molten cast iron), and hence list the complete analysis of the final (cupola's) product (molten cast iron).

Solution

For Phosphorous in Pig Iron

$$m_{P-1} = m_{MC} \cdot (\text{fraction of pig iron in the MC}) \cdot (\text{fraction of P in the pig iron})$$
$$m_{P-1} = 1000 \times 0.35 \times 0.0014 = 0.49 \text{ kg}$$

For Phosphorous in Cast Iron Scrap

$$m_{P-2} = m_{MC} \cdot (\text{fraction of cast iron scrap in the MC})$$
$$\cdot (\text{fraction of P in the cast iron scrap})$$
$$m_{P-2} = 1000 \times 0.30 \times 0.0019 = 0.57 \text{ kg}$$

For Phosphorous in Cast Iron Returns

$$m_{P-3} = m_{MC} \cdot (\text{fraction of cast iron returns in the MC})$$
$$\cdot (\text{fraction of P in the CI returns})$$
$$m_{P-3} = 1000 \times 0.35 \times 0.0015 = 0.525 \text{ kg}$$

For Phosphorous in Total Metal Charge

$$\text{Mass of phosphorous in the metal charge} = m_P = m_{P-1} + m_{P-2} + m_{P-3}$$
$$m_P = 0.49 + 0.57 + 0.525 = 1.585 \ \text{kg}$$

$$\text{Percentage of phosphorous in the final} \left(\text{molten} \right) \text{cast iron} = \frac{m_P}{m_{MC}} \times 100$$

$$= \frac{1.585}{1000} \times 100 = 0.158$$

The complete analysis of final (molten) cast iron is as follows:

$$C : 3.64 \ \text{wt.\%}, Si : 2.15 \ \text{wt.\%}, Mn : 0.49 \ \text{wt.\%}, S : 0.077 \ \text{wt.\%},$$
$$P : 0.158 \ \text{wt.\%}, Fe : \text{balance.}$$

Questions and Problems

5.1. Draw a labeled sketch of a cupola furnace showing various parts and zones.
5.2. Why does the temperature decrease in the reduction zone during the cupola operation?
5.3. List all the inputs and outputs in a cupola operation.
5.4. What are the advantages and limitations of cupola furnace?
5.5. List the cupola's design parameters and their nomenclature, and hence show them in a sketch.
5.6. Encircle the most appropriate answers for each of the following statements.

    (1) The portion of cupola above the charging door is called ______.
    (a) Stack zone, (b) melting zone, (c) reducing zone, (d) combustion zone
    (2) Tuyeres are located in ______________.
    (a) Stack zone, (b) melting zone, (c) reducing zone, (d) combustion zone
    (3) The portion of cupola between the tuyeres and the sand bottom is called the ______.
    (a) Stack, (b) wind box, (c) well, (d) melting zone
    (4) Which material is added till the level of 50 cm above the tuyeres?
    (a) Cast iron scrap, (b) coke, (c) pig iron, (d) limestone
    (5) Which material is added at the bottom of cupola?
    (a) Sand (b) coke, (c) limestone, (d) pig iron
    (6) Which part of cupola acts as a spark arrester?
    (a) Tuyeres, (b) wind box, (c) hood, (d) charging door

P-5.7. The inside diameter of a cupola at the tuyere's level is 1050 mm. The metal/coke ratio, to be used, is 18. Determine the: (a) melting rate of iron, and (b) mass of molten iron tapped for 3 taps in an hour.
P-5.8. It is recommended to use a metal/coke ratio of 18 for a cupola furnace operation. The required melting rate is 4 tons/h. The refractory brick's thick-

Table 5.6 Impurity element contents in each metal-charge constituents

Constituent	Pig iron	Cast-iron scrap	Cast-iron returns
Carbon (C)	3.8%	3.4%	3.3%
Silicon (Si)	2.5%	2.3%	2.2%
Manganese (Mn)	0.7%	0.6%	0.65%
Sulfur (S)	0.016%	0.04%	0.036%
Phosphorous (P)	0.13%	0.17%	0.16%

ness at the tuyere's level is 8 mm. The shell is constructed by 5-mm-thick steel plates. Determine the (a) inside diameter, and (b) outer diameter at the tuyere's level.

P-5.9. By using the data in P-5.8, determine the metal holding capacity of the well of the cupola. The height of tuyeres' level (from slag hole) is 300 mm. The height of the slag hole from the sand bottom is 500 mm. The fraction of the well's space to hold the metal is 40%.

P-5.10. The percentage of coke, in a cupola-furnace charge, is 15. The carbon (C) content of the coke is 90%. The percentage of carbon pick by the metal is 0.50. Determine the mass of carbon burnt (kg) per 100 kg of iron melted.

P-5.11. By using the design data in P-5.9, determine the diameter of each tuyere if the tuyeres' ratio is 10.

P-5.12. The inside diameter of cupola at the tuyeres' level is 900 mm. The density of coke is 480 kg/m^3. Determine the mass of the coke-bed charge.

P-5.13. By using the data in P-5.10, determine the blast rate for the cupola operation. The melt rate is 9 tons/h. Hint: refer to Fig. 5.4.

P-5.14. A foundry supervisor has measured the outer perimeter of a cupola at the tuyere's level as 3142 mm. The refractory brick's thickness of a cupola at the tuyere's level is 8 mm. The shell is constructed by 5-mm-thick steel plates. What is the inside diameter at the tuyere's level?

P-5.15. A cupola's metal charge has a mass of 1.2 ton. The metal charge is made up of the following constituents: pig iron (37%), cast-iron scrap (30%), and cast-iron returns (33%). The contents of impurity elements (carbon, silicon, manganese, sulfur, and phosphorous) in each constituent are presented in Table 5.6. The carbon pick up from coke is 0.15%. The silicon loss during melting is 11% of total silicon in the metal charge. The Mn loss during melting is 19% of total Mn in the metal charge. There is 5% gain in sulfur content. Predict the complete analysis by determining the contents of all elements in the final cupola product (molten cast iron).

References

Alibi, S. A., & Mahachi, J. (2022). Performance of concrete made from cupola furnace slag and recycled concrete aggregate. *MaterialsToday PROCEEDINGS, 62*(Suppl. 1), S110–S114.

ASHRAE. (1997). *ASHRAE 1997 Handbook—Fundamentals, GA. p. 3.12.* American Society of Heating Refrigeration and Air-Conditioning Engineers.

BCIRA. (1979). *Cupola design, operation and control.* British Cast Iron Research Association (BCIRA).

DoE (2012). Furnace blower fan energy consumption, *Department of Energy* (DoE). Online: https://www1.eere.energy.gov/buildings/appliance_standards/pdfs/ff_prelim_app_07_c_furnacefanconsumption_2012_06_22.pdf

Fisher, R. (2024). Metal casting, *Reliance Foundry Inc.,* Surrey, BC, Online: https://www.reliancefoundry.com/blog/castiron?srsltid=AfmBOoot5PUFt0bWtJV7PzKHJN4ik_6hqgHNx7kT4fXslE0pleyjhL3A

Huda, Z. (2020). *Metallurgy for physicists and engineers.* CRC Press.

Indian standard Institution. (1983). *Recommended sizes of cupola furnace for foundry,* Online: https://law.resource.org/pub/in/bis/S10/is.5032.1983.pdf

Junghbluth, H. (1939). Die Gesetze des Kupolofenschmelzens. *Die Giesserei, 113.*

Khatemi, B., & Longa, W. (2005). Determining optimum volume of cold and hot blast air in single-row coke cupolas. *Metallurgy and Foundry Engineering, 31*(1), 53–65.

Lanjewar, B., Jayan, N., Chaware, A., et al. (2023). Effect of cupola slag as a coarse aggregate on compressive strength of concrete. *MaterialsToday PROCEEDINGS*, In Press, Online: https://www.sciencedirect.com/science/article/abs/pii/S2214785323021107

Longa, W. (1998). Calculation of melting zone height in coke cupolas. *Archives of Metallurgy, 43*(2).

Longa, W., & Freibe, G. (1999). Wärmetheoretische Berechnungen der Vorwärmzone eines Kupolofens. *Giesserei Forschung, 51*(3).

Longa, W., & Freibe, G. (2001). Berechung der stabilen Höhe der Schmelzzone in koksbetriebenen Kupolofen. *Giesserei Forschung, 53*(2).

Nwajagu, C. O. (1994). *Foundry theory and practice.* Enugu state University.

Ocheri, C. (2020). Design, fabrication and construction of cupola furnace for metallurgical industries. *Journal of Applied Material Science & Engineering Research, 4*(4), 134–140.

OTCA. (1984). *Foundry engineering (iron foundry practice).* Overseas Technical Cooperation Agency.

Part II
Casting Design and Processes

Chapter 6
Permanent-Pattern Expendable-Mold Casting Processes

Nomenclature

P_n	Permeability number
SiO_2	Silica
M	Moisture content
m_1	Mass of original sand sample
m_2	Mass of the dried sand after reaching a constant value
A	Mass % retained on each sieve in a sieve shaker
B	The previous sieve number relative to A
m_i	Mass retained on the sieve in the set of sieves
$\sum m$	Sum of all retained masses

Abbreviations

AFS	American Foundry Society
GFN	Grain fitness number
PPEMC	Permanent-pattern expendable-mold casting
V-casting	Vacuum casting

6.1 PPEMC Processes: Definition and Classification

Permanent-pattern expendable-mold casting (PPEMC) processes refer to those metal casting processes that are based on multiple-use patterns, but single-use molds. These methods involve the use of temporary, non-reusable molds. These processes are most commonly employed in foundries. Permanent patterns, for PPEMC, may be made of wood, metal, or plastic.

PPEMC processes may be classified based on the following types of bond in the molding materials: (i) sand-, water-, and clay bond; (ii) resin bond; (iii) silicate

© The Author(s), under exclusive license to Springer Nature Switzerland AG 2025

Z. Huda, *Metal Casting Engineering*, Mechanical Engineering Series, https://doi.org/10.1007/978-3-031-84620-5_6

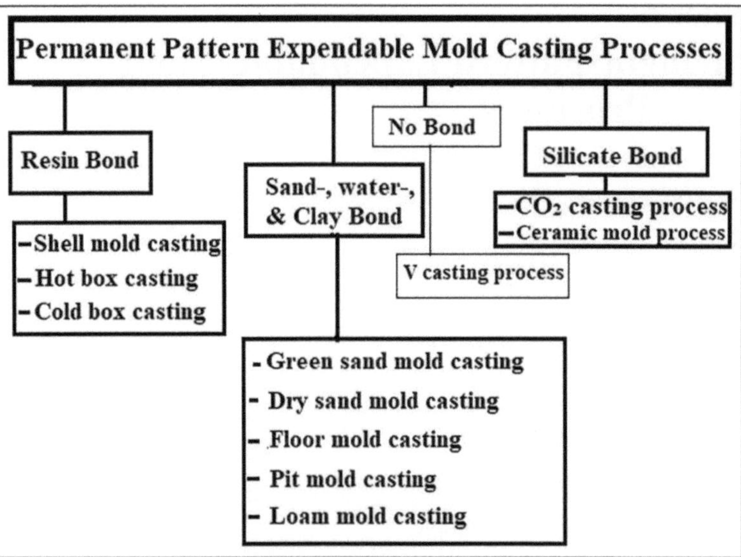

Fig. 6.1 Classification of PPEMC processes

bond; and (iv) no bond processes (see Fig. 6.1). The examples of sand, water, and clay bond PPEMC processes include green sand mold casting, dry sand mold casting, floor mold casting, pit mold casting, and loam mold casting. These processes are discussed in the following sections. The design of sand casting is discussed in Chap. 7.

6.2 Properties of Foundry Sand

The sand-, water-, and clay bond PPEMC processes are generally known as sand casting; however, there are other sand-water-clay based processes, such as floor and pit mold casting, loam mold casting, etc. In general, the sand (SiO_2) is held together by a mixture of water and clay; the sand-water-clay mixture is called *green sand*. The control of all aspects of the properties of sand is important when manufacturing parts by sand casting. This is why a sand laboratory is usually attached to the foundry. Molding sand has three important advantages over other molding materials: (a) it is inexpensive, (b) it can be easily recycled, and (c) it withstands extremely high temperatures.

Foundry/molding sand must possess the following properties: (1) strength, (2) permeability, (3) moisture content control, (4) flowability, (5) grain size control, (6) adequate grain shape, (7) collapsibility, (8) refractoriness, and (9) re-usability (Fisher, 2024). These properties are discussed in the following paragraphs.

Strength of foundry sand refers to its ability to hold its geometric shape under the conditions of mechanical stress in a sand mold. The sand should have sufficient compressive strength so that it can easily retain its shape during conveying, turning, closing, and pouring. Sand with a low compression strength leads to casting defects. The strength of green sand is called *green strength*. Angular-shaped grains of sand impart good bonding strength.

Permeability refers to the ability of a sand mold to permit the escape of gases and steam during the casting process. It can be experimentally measured using a standard AFS sample (50.8 mm in diameter and 50.5 mm tall) by passing 2000 cm^3 of air at a gauge pressure of 10 g/cm^2. The time required for exactly 2000 cm^3 of air to pass through the sample is recorded by a stop watch. The absolute permeability number (P_n) can be calculated by:

$$P_n = \frac{V \bullet h}{p \bullet A \bullet t} \tag{6.1}$$

where V is the volume of air passing through the sample (V = 2000 cm^3), h is the height of the sample ($h = 5.05$ cm), A is the cross-sectional area of the sample (cm^2), p is the pressure of air in cm of water, and t is the time (min) (see Example 6.1).

Moisture content in molding sand greatly influences a mold's strength and permeability; a mold with too little moisture may collapse, while a mold with too much moisture can result in the entrapment of steam bubbles in the casting. In order to experimentally determine the moisture content, a 50 g-sample of green sand is dried in an oven at 105 ± 5 °C. The sample is cooled in a desiccator to room temperature and then the dried sample is weighed. The process is repeated until the mass is found constant. *Green sand moisture content* is defined as the amount of free water contained in the green sand. It can be calculated by (Nakayama Co., 2024):

$$\%\mathrm{M} = \frac{m_1 - m_2}{m_1} \times 100 \tag{6.2}$$

where $\% M$ is the percentage of moisture content in the green sand, m_1 is the mass of the original sample (g), and m_2 is the mass of the dried sand after reaching a constant value (g) (see Example 6.2). In general, green sand moisture content is in the range of 6–10 wt.%.

Flowability refers to the ability of the molding sand to fill small cavities in the pattern. High flowability results in a more precise mold, and is, therefore, useful for intricate castings. Rounded-shape grains of sand correspond to better flowability.

Grain size is the average size of sand particles. The size of grains or particles and their distribution is determined by use of sieve shakers, which enables us to determine the grain fitness number (GFN). A typical sieve shaker separates sand particles by passing them through a series of chambers with mesh filters and agitating the sample thereof (see Fig. 6.2).

The American Foundry Society grain fineness number (AFS-GFN) is a measure of the average size of the grains (or particles) and their distribution in a sand system.

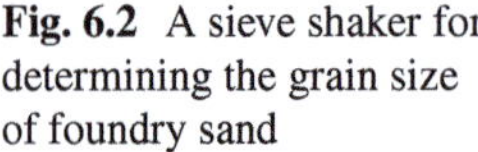

Fig. 6.2 A sieve shaker for determining the grain size of foundry sand

The optimal GFN in a sand system is determined by the type of metal poured, pouring temperature, required surface finish, and the like. Sand that is too fine (high GFN) can create low permeability issues. On the other hand, sand that is too coarse (low GFN) can create high permeability problems.

The determination of GFN, by use of a sieve shaker, involves the following procedure: (I) cleaning the meshes of each sieve with the brush provided, (II) placing the required set of standard sieves with the coarsest mesh on the top and finest at the bottom, (III) placing the required mass quantity of washed/dried sand at the top sieve, (IV) putting the lid on the top sieve, (V) transferring the nest of sieves on the shaker and clamping them in the position, (VI) turning on the motor to shake the sieves, (VII) stopping the motor after required time and removing the sieves, (VIII) carefully removing the lid from the top sieve and separating each sieve, (IX) weighing each sieve and then finding the mass retained on each sieve, and (X) checking sieve number and mesh size and finally recording them in the form of a table.

By using the data for the mass(es) retained on each sieve, the mass % retained on each sieve can be calculated by:

$$\text{Mass\% retained on each sieve} = A = \frac{m_i}{\sum m} \times 100 \tag{6.3}$$

where m_i is the mass retained on the sieve in the set of sieves (g), and $\sum m$ is the sum of all retained masses (g). The AFS-GFN is determined by the following formula:

$$\text{AFS} - \text{GFN} = \frac{\sum(A \cdot B)}{\sum A} = \frac{\sum(A \cdot B)}{100} \tag{6.4}$$

where B is the previous sieve number relative to A, and $\sum(A \bullet B)$ is the sum of products of A and B. The significance of Eqs. 6.3 and 6.4 is illustrated in Examples 6.3, 6.4, and 6.5.

Grain shape refers to the shape of the individual grains of sand based on how round they are. Strictly speaking, there are six (6) types of grain shapes in molding sands: (a) well-rounded, (b) rounded, (c) sub-rounded, (d) subangular, (e) angular, and (f) very angular (see Fig. 6.3). Molding sand with rounded grains provides relatively poor bonding strength, but good flowability and surface finish. Angular grains impart greater bonding strength because of interlocking, but they result in poorer flowability and permeability than rounded-grain sands. Subangular grains have intermediate shape between round and angular grains. They possess better strength and lower permeability as compared to rounded grains, but lower strength and better permeability than angular grains.

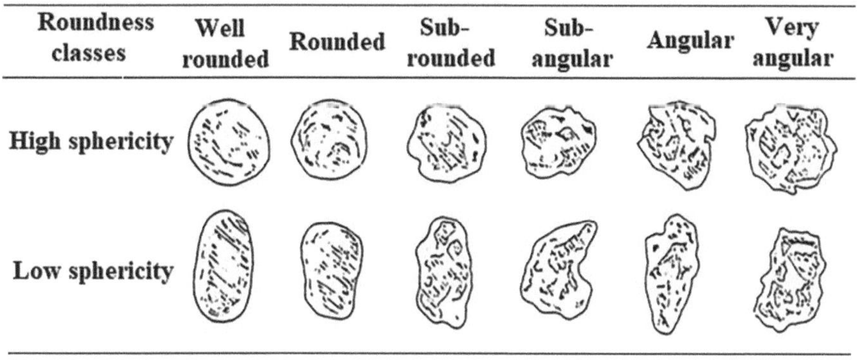

Roundness classes	Well rounded	Rounded	Sub-rounded	Sub-angular	Angular	Very angular
High sphericity						
Low sphericity						

Fig. 6.3 Grain shapes in molding sand

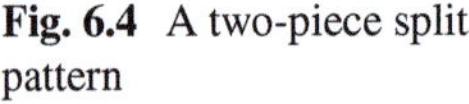

Fig. 6.4 A two-piece split pattern

6.3 Pattern Making

6.3.1 What Is a Pattern in Metal Casting?

The first step in all PPEMC processes is to make a permanent pattern with the same shape and geometry as desired in the product (casting). A pattern is a model for the object to be cast; it makes an impression (cavity) in the mold into which liquid metal is poured. The pattern may be made of wood, plastic, or metal. There are several types of patterns, which include a two-piece split pattern (Fig. 6.4), one-piece pattern, match-plate pattern, cope-and-drag pattern, etc.

6.3.2 Allowances in Pattern Design

A pattern should be slightly oversized (as compared to the casting) to allow the following allowances: (a) shrinkage allowance, (b) draft allowance (see Fig. 6.5), (c) machining allowance, and (d) distortion allowance. Wooden patterns are generally made by carpenters. However, the recent trend is to make polymeric patterns by additive manufacturing—3D printing. A detailed account of pattern design is given in Chap. 7 (Sect. 7.1).

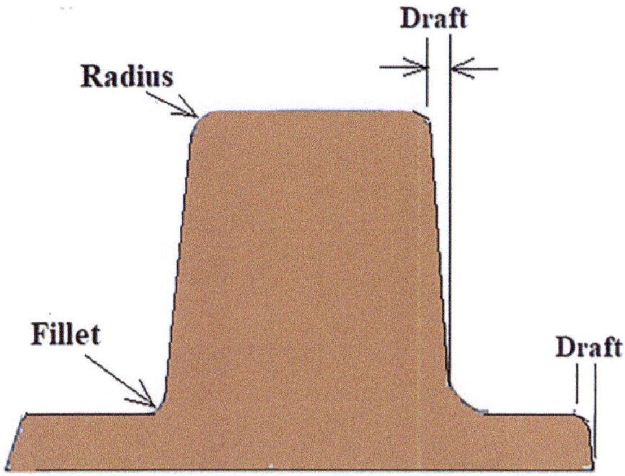

Fig. 6.5 A pattern with draft allowance

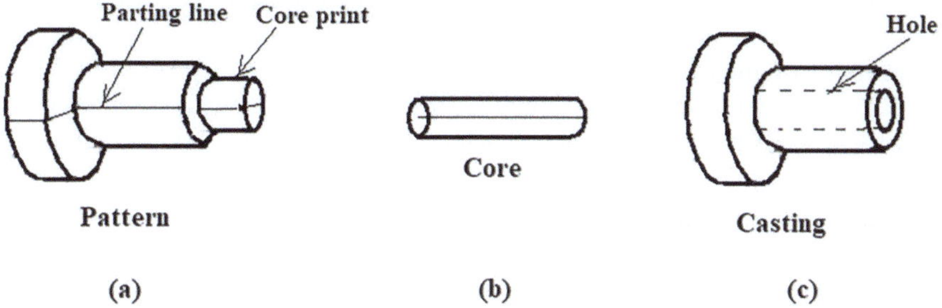

Fig. 6.6 Pattern with core print (**a**), core (**b**), and casting with a hole (**c**)

6.3.3   Patterns with Core Prints: Core and Core Print

A *core print* is the recess added to a pattern (see Fig. 6.17a). The purpose of a core print is to provide a seat for the core and to support it while the metal is being poured. A *core* is a piece of hardened sand (Fig. 6.6b); it is used to create holes and various shaped cavities in the castings (Fig. 6.6c). In industrial practice, many machine components require holes or cavities in their design. These castings are produced by use of patterns with core prints.

6.4   Green Sand Mold Casting Process: Sand Casting Process

6.4.1   Green Sand and Foundry Tools

Strictly speaking, green sand, for metal casting, consists of high-quality silica sand, bentonite clay (as the binder) ($\approx$10%), water (2–5%), and sea coal ($\approx$5%) (carbonaceous mold additive to improve casting finish) (F.H.A, 2016). Shilpa and

Fig. 6.7 One-half pattern is placed in one of the two flasks

co-researchers (2021) have experimentally investigated the effects of various ingredients of green sand; they have reported that elevated bentonite and coal powder in the sand could prominently improve the shear strength, compression strength, and permeability of green sand (Shilpa et al. 2021).

Mold making, for sand casting, requires the use of the following essential tools: (a) *flask or mold box*—two metallic frames into which green sand is filled; upper flask is called the cope, and the lower flask is drag, (b) *draw spike*—a tool used to remove pattern from sand mold, (c) *strike-off bar*—a tool used to strike off excess sand from the top of a molding box, (d) *rammer*—the sand compactor, (e) *sprue and riser*—cylindrical pieces used to make sprue hole and riser hole in the sand mold, (f) *trowel*—tool for digging to make runner for flow of metal, and (g) bottle containing parting compound (e.g., graphite powder) to facilitate easy removal of pattern from the mold.

6.4.2 Steps in Sand Casting Process

It has been learnt in Chap. 1 that the sand casting process involves the following principal steps: (a) preparation of pattern and sand mold, (b) melting of metal, (c) pouring the molten metal into the mold, (d) allowing the metal to cool/solidify in the mold, (e) removing the casting from the mold, and (f) cleaning and quality control.

The following steps must be followed to prepare a sand mold. (I) One-half of the pattern is placed on a wooden board (flat surface), and one of the flask (cope) is placed around the half-pattern (Fig. 6.7). (II) Parting compound (graphite powder) is sprayed onto the half-pattern and sand is filled into the flask (cope) (Fig. 6.8). (III) The sand, in the cope, is rammed to ensure that the sand is securely holding the pattern (Fig. 6.9). (IV) The excess sand is scrapped-off by using a strike-off bar (Fig. 6.10). (V) The cope is flipped over, and the other half pattern is placed over the

Fig. 6.8 Green sand is filled into the flask containing half-pattern

Fig. 6.9 Ramming of sand in the flask

first half to complete the pattern; the other flask (drag) is placed on the cope (Fig. 6.11). (VI) A sprue is designed (see Chap. 7) and erected (positioned) in the drag (Fig. 6.12). (VII) A riser is designed (see Chap. 7) and positioned in the drag, as shown in Fig. 6.13. (VIII) The sprue and riser are carefully removed to create sprue hole and riser hole, respectively; thus, a partial gating system is formed. (IX) The two halves of the pattern are removed by use of a draw spike (Fig. 6.14). (X) A runner (passage way to allow flow of molten metal from the sprue hole to the mold cavity) is created by use of a trowel (Fig. 6.15), thereby completing the gating

Fig. 6.10 Scrapping off
the sand by use of a
strike bar

Fig. 6.11 The two halves
of pattern are together; the
drag is placed on the cope

system. (XI) The cope and drag are assembled, and vents are provided in the assembled mold; now, the sand mold is ready for pouring of molten metal (see Fig. 6.16).

Once the sand mold has been prepared, the molten metal is poured into the mold (see Chap. 4, Fig. 4.5b). It is important to allow the metal to cool in the mold for the predetermined solidification time (see Chap. 3, Sect. 3.6). The next step is to remove the casting and clean it (see Fig. 6.17).

Unless well-designed, sand castings have defects. For example, residual stress is a serious defect in sand mold casting, such as that used to manufacture cylinder heads and machine-tool beds. In order to predict this defect, a finite element method (FEM) thermal stress analysis should be conducted (Yuki, et al., 2013). The techniques to avoid casting defects are discussed in detail in Chap. 12.

Fig. 6.12 The drag (with pattern and sprue) placed on the cope

Fig. 6.13 The drag (with pattern, sand, sprue, and riser) placed on the cope followed by filling of sand in the drag

6.5 Other Sand-Water-Clay Bond Casting Processes

It has been learnt in Sect. 6.1 that the examples of sand, water, and clay bond PPEMC processes include green sand casting, dry sand casting, floor and pit mold casting, and loam mold casting.

Fig. 6.14 Use of draw
spike to remove a half
pattern

Fig. 6.15 Mold cavity
(half) with runner

6.5.1   Dry Sand Casting

Dry sand molding or dry sand casting is a specialized form of green sand casting
process. It involves baking a sand mold at a given temperature to make the mold
stronger. This process is generally used in large foundries to manufacture big ferrous and non-ferrous castings (e.g., engine blocks, construction parts, etc.).

The following are the distinct advantages of dry sand casting: (a) heavy and
accurate castings can be produced; (b) the moisture content in the mold is negligible, thereby increasing its compressive strength; (c) this process is highly suitable
for producing heavy casting with close tolerance, surface finish, and dimensional
accuracy; (d) there is high flexibility in adding chemical bonds and additives to the

Fig. 6.16 Sand mold is ready for molten-metal pouring

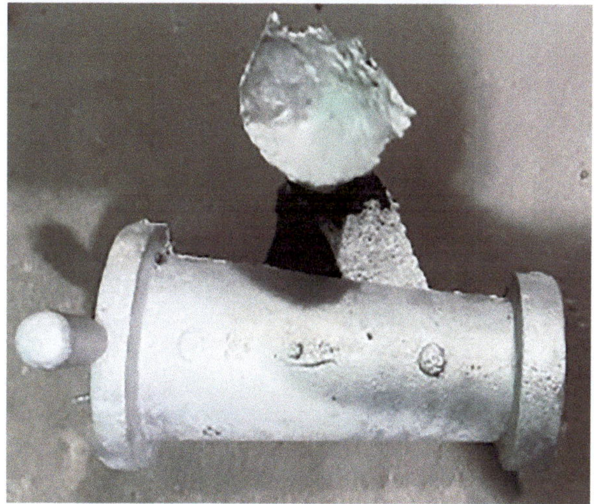

Fig. 6.17 The product—casting (the attached riser and gating must be removed)

molding sand; and (e) castings produced by this process have negligible gas-related defects, such as porosity, blowholes, and pin-holes.

6.5.2 Floor Mold Casting

Floor molding or floor mold casting involves the preparation of sand mold on the floor of the foundry. In this process, the molding is done on the floor level itself. In dry sand molding, foundry floor sand bed is used as a drag. A two-piece split pattern is used; one-half of the pattern is fully embedded in the floor sand and its top face is leveled by compacting sand around the pattern. The other half of the pattern is, then, placed over the sand-embedded half-pattern and the assembly enclosed with a flask (cope). Sand is filled in the cope.

Molds for medium and large-sized castings are prepared by floor molding because large molds are difficult to handle. Since the mold cannot be shifted, the metal is directly poured into the mold cavity of the floor.

6.5.3 Pit Mold Casting

Pit mold casting or *pit molding* involves digging a pit in the foundry floor. Since very large castings cannot be accommodated in mold boxes, they are molded in pits. The pit is lined with refractory bricks; the desired shape of the mold is given by using a pattern. The pit bottom carries a well-rammed layer of cinder to allow escape of gases during casting. The following steps are involved in pit molding. (I) The upper flask (cope) is placed over the pit containing the pattern. (II) The cope is filled with sand and compacted and later provision of sprue and riser is made in the cope sand. (III) Once the cope has been lifted, the pattern is removed from the sand bed (housed in the pit). (IV) Finally, the cope is placed back in position in the pit to make the mold ready for pouring.

6.5.4 Loam Mold Casting

Loam mold casting process is used for manufacturing extremely large castings that are symmetrical. In this process, firstly, a rough skeleton of the mold is made using bricks reinforced with iron plates. Then, loam sand is daubed over and plastered on the brick skeleton to make the mold face, which is later shaped to size. A refractory facing is then given to the mold face and the mold dried to increase its strength. A loam sand mortar is prepared using coarse silica sand, clay, chopped straws, and fire-clay milled with water. Loam mold casting is best suited for producing large-sized cylinders, kettles, gear blanks, and the like.

6.6 Resin Bond Casting Processes

6.6.1 Shell Mold Casting

6.6.1.1 Advantages and Applications

Shell mold casting, or shell molding is permanent-pattern expendable-mold casting (PPEMC) process that uses a resin-bonded sand to form the mold. This process has the following advantages: (a) better dimensional accuracy can be achieved, (b) the

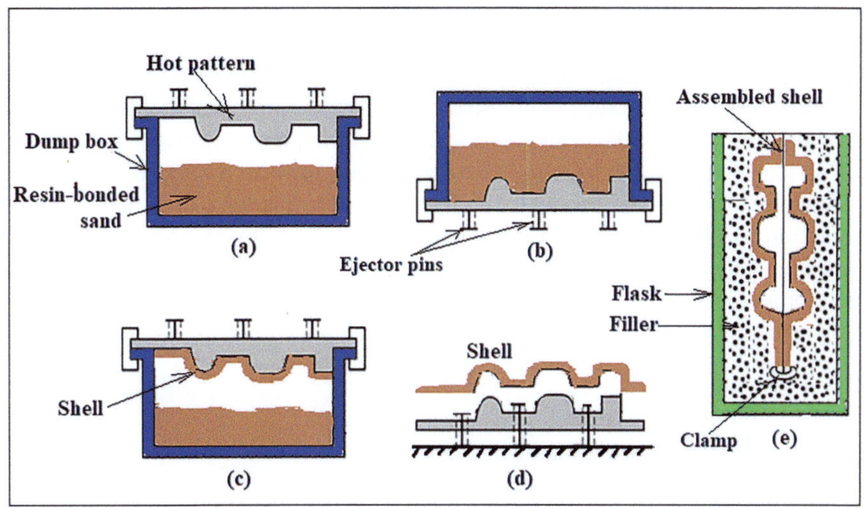

Fig. 6.18 Steps in shell molding

productivity rate is higher, and (c) labor requirements are lower. Shell mold casting process is well suited for manufacturing automotive gear housings, cylinder heads, and connecting rods, and the like.

6.6.1.2 Steps in Shell Mold Casting

The following are the overall steps in shell molding: (a) pattern making—a two-piece metal pattern is made in the shape and geometry of the desired product, (b) mold/shell making, (c) mold/shell assembly, (d) pouring of molten metal, and (e) solidification and casting removal.

Mold/shell preparation involves the following steps. (I) Each half-pattern is heated to a temperature in the range of 175–370 °C and coated with a lubricant to facilitate its removal. (II) The hot pattern is clamped to a dump box, which contains the resin-bonded sand (see Fig. 6.18a). (III) The dump box is vertically flipped (turned upside down), thereby allowing the sand-resin mixture to coat the hot pattern (Fig. 6.18b). The hot pattern partially cures the mixture forming a shell around the pattern. (IV) The dump box is again vertically flipped (Fig. 6.18c). (V) The shell is ejected from the pattern by use of ejector pins (Fig. 6.18d). (VI) The two halves of the shell are assembled and clamped in a flask that is filled with a filler material (see Fig. 6.18e).

6.6.1.3 Recent Development in Shell Molding

Although the shell mold casting process is a resin bond PPEMC process, recent research has shown the successful use of sand-water-clay in shell molding, which is explained as follows. Hatsu and co-researchers (2021) have investigated the effects of various combinations of clay, silica, and water to form a mold for shell-mold casting of a machine component. Of all the mixtures that were fired, the best mixture ratio was found to be the sample with the composition: 70% Mfensi clay, 10% silica sand, and 20% water. The shell core, made in the experiment, exhibited better core properties, such as collapsibility, thermal stability, and permeability when the molten cast iron was poured into the mold for solidification (Hatsu et al., 2021).

6.6.2 *Hot-Box Casting Process*

A *core* is generally baked in a *core box* (Fig. 6.19), which may be either a hot box or cold box. The hot-box casting is the process for making cores out of moist, organically bound mold materials that are hardened by heating in hot-metal core boxes. The stored (absorbed) heat in the core making process results in full hardening of the core. This process enables the production of cores of high quality and high complexity due to excellent shaping accuracy.

The hot box casting process involves the following steps: (I) preparation of the mixture (sand + resin binder + catalyst), (II) transportation of the material to the core shooting machine, (III) injection of the mixture into the preheated (180–250 °C) core box using heaters, the shooting pressure should be around 6 atm., (IV) hardening of the sand mixture in contact with the hot mold, and (V) removal of the core and post-hardening of the central part. After hardening, the cores must be held for about 2½ h to avoid the subsequent formation of gas-holes during casting.

6.6.3 *Cold Box Casting*

The cold-box casting process involves sand core production based on a chemical reaction between sand and resins at room temperature. The reaction is accelerated by a catalyst. The advantages of this process include: (a) high productivity, (b) the

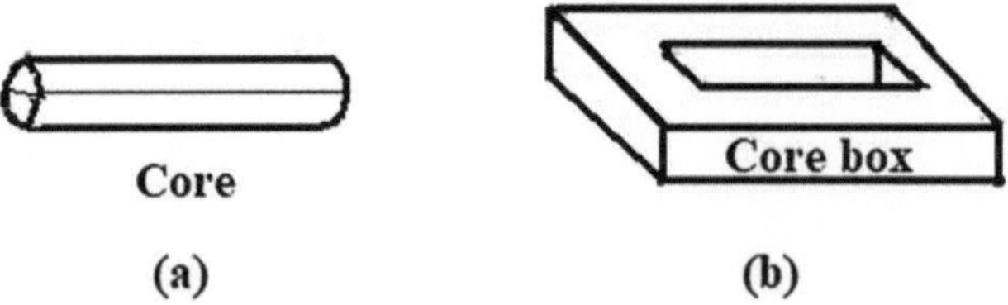

Fig. 6.19 A core (**a**), and a core box (**b**)

ability to cast thin walled sections and complex shapes, (c) no heating required in the core box, and the like.

In the cold box casting process, the sand-resin mixture is shot into the mold and passed through a catalyst. The components (resin and catalyst) are mixed with the sand in a discontinuous machine, and then shot into the casting system. After injection into the mold, a tertiary amine (e.g. DMEA, DMPA, TEA, etc.) is added and, finally, the catalyst is removed.

6.7   Silicate Bond Casting Processes

6.7.1 CO_2 Casting Process

The CO_2 casting process is basically a sand casting process in which the mold is hardened by blowing carbon dioxide (CO_2) gas over it (see Fig. 6.20). The mold is made of a mixture of sand and silicate binder. It is hardened by blowing CO_2 gas using a CO_2 cylinder, regulator, hoses and nozzle. This process allows for dimensionally accurate castings with fine surfaces at a lower cost than those produced by the green sand casting process.

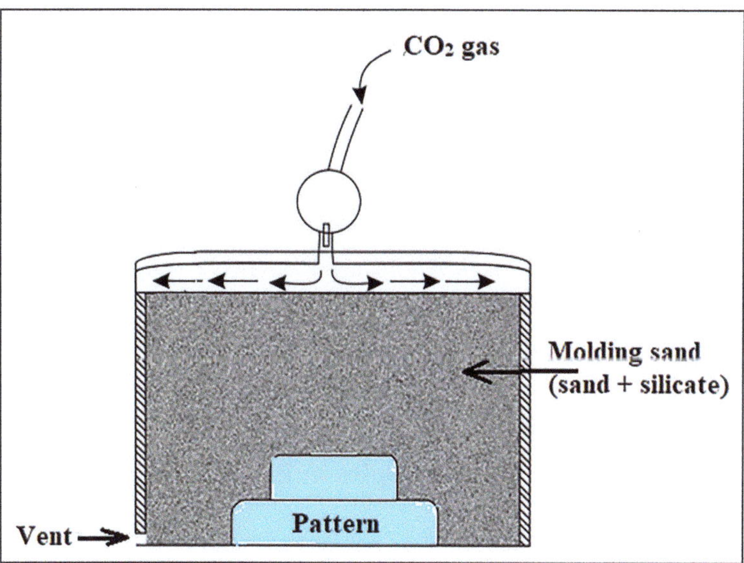

Fig. 6.20 Schematic of CO_2 casting process

6.7.2 Ceramic Mold Casting Process

The ceramic mold casting is a PPEMC process that involves the combination of materials to make a ceramic slurry, which has a rubber-like consistency. In this process, reusable and cheap patterns made of wood, metals, plastic, or rubber are used; patterns with intricate designs can be used since the molding material supports castings of complicated designs.

There are two types of ceramic mold casting process: (a) true ceramic mold casting, and (b) Shaw process. In the case of true ceramic mold casting, the slurry can be made of a mixture of zinc and calcined high-alumina mullite. In the Shaw process, the molding mixture consists of refractory filler + hydrolyzed ethyl silicate + a liquid catalyst. In both cases, the patterns are kept in a flask and the slurry is poured over the pattern. The molding material sets in a rubber-like consistency around the pattern, which makes it easy for the pattern to be stripped out of the mold. The mold, which is made in cope and drag form, is heated to harden it. The strong mold is useful for high temperature pours, which can be done when the mold is still hot.

Ceramic mold casting process is best suited for producing stainless steel and bronze products ranging from household goods to industrial tools. Some examples of cast products include kitchenware (e.g., kettles) and industrial products (e.g., impellers, complex cutting tools, plastic molding tooling, etc.).

6.8 V-Casting Process

The V-casting process is an advanced form of sand casting process that utilizes very fine dry sand with no binder. In this process, sand is held in shape by vacuum packing it between sheets of thin plastic film. This vacuum causes the sand to become hard. Then the pattern is removed, and molten metal is poured into the cavity. Finally, the vacuum is released, and the casting is removed. The advantages of V-cast products include: (a) thinner walls, (b) cosmetically smooth surface finishes, (c) tighter tolerances, and (d) vertical walls because the V-process requires no draft angles.

6.9 Worked Numerical Examples in PPEMC Processes

Example 6.1 Determination of the Absolute Permeability Number of Molding Sand

The permeability of molding sand was experimentally determined using a standard AFS sample by passing 2000 cm^3 of air at a gauge pressure of 10 g/cm^2. The time taken for the air to escape was 1.3 min. Determine the absolute permeability number.

Solution

$V = 2000$ cm³, $h = 5.05$ cm, D $= 5.08$ cm, p $= 10$ g/cm², $t = 1.3$ min.

$$A = \frac{\pi}{4} \cdot D^2 = \frac{\pi}{4} \times 5.08^2 = 20.268 \, cm^2$$

By using Eq. 6.1,

$$P_n = \frac{V \cdot h}{p \cdot A \cdot t} = \frac{2000 \times 5.05}{10 \times 20.268 \times 1.3} = 38.33$$

The absolute permeability number $= 38.33$.

Example 6.2 Determination of the Green Sand Moisture Content

A standard experiment to determine the green sand moisture content yielded the mass of the dried sample as 46 g. Determine the green sand moisture content.

Solution

$m_1 = 50$ g, $m_2 = 46$ g, M $=?$

By using Eq. 6.2,

$$\%M = \frac{m_1 - m_2}{m_1} \times 100 = \frac{50 - 46}{50} \times 100 = 8$$

The green sand moisture content $= 8\%$.

Example 6.3 Determination of Mass % Retained on a Sieve in a Sieve Shaker Test

A foundry-sand system was experimentally tested by use of a sieve shaker. The mass retained on a sieve (in the set of sieves) is 21 g. The original mass of the sand taken is 300 g. Determine the mass % retained on the sieve.

Solution

$m_i = 21$ g, $\sum m = 300$ g.

By using Eq. 6.3,

$$Mass\%\,retained\,on\,each\,sieve = A = \frac{m_i}{\sum m} \times 100 = \frac{21}{300} \times 100 = 7$$

The mass retained on the sieve $= 7\%$.

Example 6.4 Completing the Sand Grain Size Distribution Data Table for a Sieve Shaker Test

A foundry-sand system was experimentally tested by use of a sieve shaker. The original mass of the sand taken is 300 g. The final data was recorded as shown in Table 6.1. Complete the table by filling the data in the empty columns.

Solution

By using Eq. 6.3, the completed table is presented as Table 6.2.

Table 6.1 The data for Example 6.4

Sieve no.	Mass retained on each sieve (g)	% retained on each sieve (A)	Previous sieve number (B)	(A • B)
20	0			
25	0			
80	210			
100	39			
120	21			
150	18			
170	0			
200	0			
Pan	12			
Sum	300			

Table 6.2 The completed data for Solution to Example 6.4

Sieve no.	Mass retained on each sieve (g)	% retained on each sieve (A)	Previous sieve number (B)	(A * B)
20	0	0	0	0
25	0	0	20	0
80	210	70	25	1750
100	39	13	80	1040
120	21	7	100	700
150	18	6	120	720
170	0	0	150	0
200	0	0	170	0
Pan	12	4	200	800
Sum	300	100	–	5010

Example 6.5 Determination of AFS Grain Fitness Number (AFS-GFN)

By using the data in Table 6.2, determine the AFS-GFN for the sand system.

Solution

$\sum(A \bullet B) = 5010$, $\sum A = 100$, GFN =?

By using Eq. 6.4,

$$\text{AFS-GFN} = \frac{\sum(A \cdot B)}{\sum A} = \frac{\sum(A \cdot B)}{100} = \frac{5010}{100} = 50.1$$

The AFS grain fitness number for the tested sand system = 50.1.

Questions and Problems

6.1. Encircle the most appropriate answers for each of the following statements.

    (1) Which type of PPEM casting process, in general, is shell mold casting?

(a) Sand-, water-, and clay bond, (b) resin bond, (c) silicate bond, and (d) no bond

(2) Which type of PPEM casting process is loam mold casting?
   (a) Sand-, water-, and clay bond, (b) resin bond, (c) silicate bond, and (d) no bond

(3) Which type of PPEM casting process is V casting?
   (a) Sand-, water-, and clay bond, (b) resin bond, (c) silicate bond, and (d) no bond

(4) Which type of PPEM casting process is CO_2 casting?
   (a) Sand-, water-, and clay bond, (b) resin bond, (c) silicate bond, and (d) no bond

(5) The recess added to a pattern for creating a hole in the casting, is called
   ___________.
   (a) Pattern allowance, (b) core print, (c) core hole, (d) core cavity

(6) Which one is the suitable material to facilitate easy removal of pattern from the mold?
   (a) Graphite powder, (b) metal powder, (c) clay powder, (d) binder powder

(7) Which PPEM casting process involves pouring of slurry over the pattern?
   (a) Pit mold casting, (b) green sand casting, (c) v-casting, (d) ceramic mold casting

(8) Which PPEM casting process involves baking of sand mold?
   (a) Pit mold casting, (b) green sand casting, (c) dry sand casting, (d) ceramic mold casting

(9) Which one is the silicate bond PPEM casting process?
   (a) CO_2 casting, (b) green sand casting, (c) shell mold casting, (d) V-casting

(10) Which grain shape of foundry sand results in the best bonding strength?
   (a) Well rounded, (b) rounded, (c) sub-rounded, (d) angular

6.2. Draw the classification chart showing various PPEMC processes.

6.3. List the properties required in foundry sand.

6.4. What is the procedure for the determination of GFN by use of a sieve shaker?

6.5. Describe the steps that must be followed to prepare a (green) sand mold.

6.6. (a) Define dry sand casting process.

(b) What are the advantages of dry sand casting as compared to green sand casting?

6.7. Diagrammatically illustrate the shell mold casting process.

6.8. Explain CO_2 casting process with the aid of a sketch.

6.9. Describe hot box casting process.

P-6.10. The permeability of molding sand was experimentally determined using a standard AFS sample by passing 2000 cm^3 of air at a gauge pressure of 10 g/cm^2. The time taken for the air to escape was 1.1 min. Determine the absolute permeability number.

Table 6.3 The data for P-6.13

Sieve no.	Mass retained on each sieve (g)	% retained on each sieve (A)	Previous sieve number (B)	(A • B)
20	0			
25	0			
80	267.5			
100	31			
120	17.5			
150	10.5			
170	10			
200	0			
Pan	13.5			
Sum	350			

P-6.11. A standard experiment to determine the green sand moisture content yielded the mass of the dried sample as 43 g. What is the green sand moisture content?

P-6.12. A foundry-sand system was experimentally tested by use of a sieve shaker. The mass retained on a sieve (in the set of sieves) is 10.5 g. The original mass of the sand taken is 350 g. Determine the mass % retained on the sieve.

P-6.13. A foundry-sand system was experimentally tested by use of a sieve shaker. The original mass of the sand taken is 350 g. The final data was recorded as shown in Table 6.3. Determine the AFS-GFN for the sand system.

References

F.H.A (Federal Highway Administration). (2016). *User guidelines for waste and byproduct materials in pavement construction*. Federal Highway Administration, US Department of Transportation. https://www.fhwa.dot.gov/publications/research/infrastructure/structures/97148/fs1.cfm

Fisher, R. (2024). *Metal casting*. Reliance Foundry Co. Ltd., USA. https://www.reliance-foundry.com/blog/sand-casting?srsltid=AfmBOoqj6nOVITXTJ5fa5k6SEpQbTWhUeFh102r1YNmFm-HdTyfPavrj

Hatsu, J. K., Sunnu, A. K., Ayetor, G. K., et al. (2021). Investigation of shell mold casting technique in Ghana using indigenous materials. *Scientific African, 14*, e01052.

Nakayama Co. (2024). *TJFS-201 Green sand moisture content test method*. Nakayama Co. Ltd, Japan. https://nakayama-meps.co.jp/english/method/TJFS-201.html

Shilpa, M., Prakash, G. S., & Shivakumar, M. R. (2021). A combinatorial approach to optimize the properties of green sand used in casting mould. *MaterialsToday: PROCEEDINGS, 39*(Pt. 4), 1509–1514.

Yuki, I., Motoyama, Y., Takahashi, H., et al. (2013). Effect of sand mold models on the simulated mold restraint force and the contraction of the casting during cooling in green sand molds. *Journal of Materials Processing Technology, 213*(7), 1157–1165.

Chapter 7
Sand Casting Design

Nomenclature

R_e	Reynolds number
Q	Volume flow rate
v_1	Flow velocity at the sprue top
v_2	Flow velocity at the sprue bottom
A_1	Cross-sectional areas at the top of the sprue
A_2	Cross-sectional areas at the bottom of the sprue
T_{MF}	Time to fill the mold cavity
V_c	Volume of casting
A_c	Surface area of casting
V_r	Volume of riser
A_r	Surface area of riser
M_c	Modulus of casting
M_r	Modulus of riser
F_b	Thrust or net buoyant force
W_m	Weight of the molten metal displaced by the core
W_c	Weight of the core
V	Volume of the core
n_b^{ch}	Number of chaplets required above the core to resist buoyancy
n_w^{ch}	Number of chaplets required beneath the core to resist weight
F_{ch}	Force sustained by one chaplet
F_m	Metallostatic force
A_p	Projection area of the casting in the parting plane
ρ_m	Density of the metal
H	Metal head ($H = h - c$)
h	Cope height = sprue height
c	Height of pattern (from the parting line) in the cope
W_s	Weight of sand in the cope
V_s	Volume of sand in the cope
ρ_s	Density of sand

© The Author(s), under exclusive license to Springer Nature
Switzerland AG 2025

Z. Huda, *Metal Casting Engineering*, Mechanical Engineering Series,
https://doi.org/10.1007/978-3-031-84620-5_7

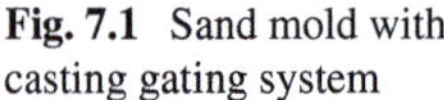

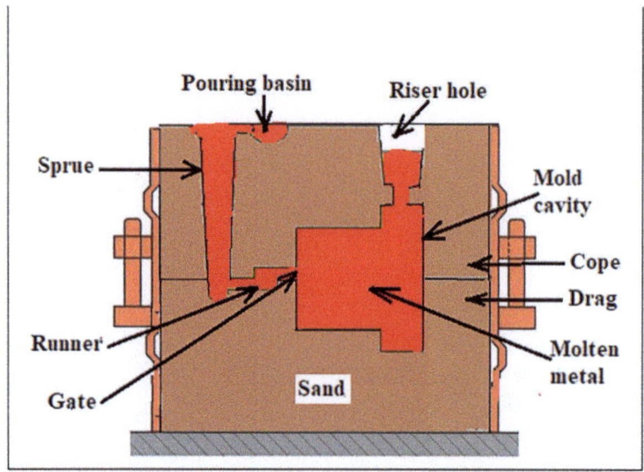

Fig. 7.1 Sand mold with casting gating system

7.1 Why Is Sand Casting Design Important?

The sand casting process has been explained in detail in the preceding chapter, and it was indicated that a poorly designed casting mold and its gating system would result in casting defects (see Sect. 6.4). It is therefore technologically important to properly analyze and design the sand casting system prior to starting the casting process. The design of a sand casting system includes pattern design, gating system design (sprue design, runner design, riser design), mold design, and the like (see Fig. 7.1). These aspects of sand casting design are discussed in the following sections.

7.2 Pattern Design

7.2.1 Patten Allowances and Parting Line

It has been learnt in the preceding chapter that a pattern should be oversized, and that the following allowances must be taken into consideration in the design of a pattern: (a) machining allowance, (b) shrinkage allowance, and (c) draft allowance. Besides these allowances, a pattern design engineer must locate the parting line and adjust the component within the mold; s/he should preferably make the parting line at mid-height of the casting, and make the line straight (see Fig. 7.2). The pattern allowances are explained one-by-one in the following sections.

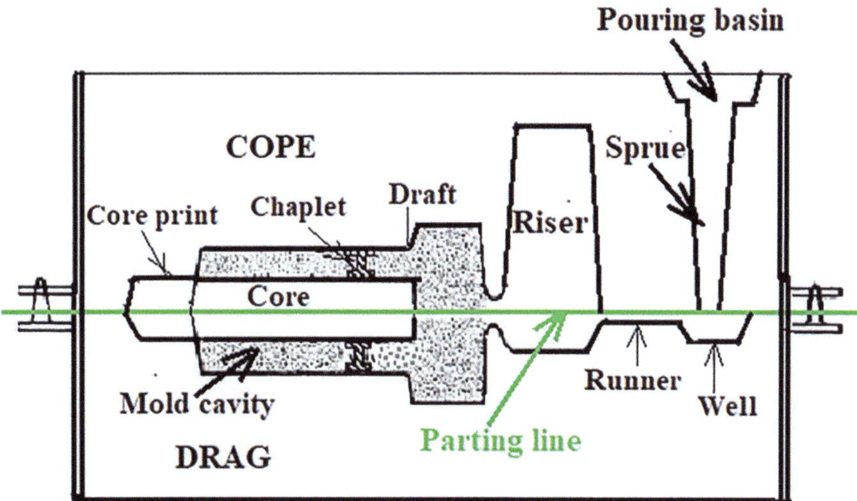

Fig. 7.2 Sand mold with core showing the straight parting line at the mid-height

7.2.2 *Machining Allowance in the Pattern*

The machining allowance, also known as the finishing allowance, is the allowance to enable the machining of the casting to the required size, accuracy, and surface finish. The machining allowance depends on the pattern size and the type of metal that is being cast. The recommended machining allowances for various pattern-sized ranges and metals are presented in Table 7.1. The significance of the data in Table 7.1 is illustrated in Examples 7.1 and 7.2.

7.2.3 *Shrinkage Allowance in Patterns:* Recent Development in Measurement

Shrinkage is the reduction in volume and size of a cast alloy, as it cools and solidifies. This shrinkage can cause several defects, particularly the shrinkage-cavity defect. It is, therefore, important to add the *shrinkage allowance* to the dimension of the desired casting to obtain pattern dimension. The linear shrinkage values for various metals and alloys are presented in Table 7.2. The significance of the data in Table 7.2 is illustrated in Example 7.3.

Figure 7.3 shows the 3D sketches of a cast-iron component (to be sand cast) with dimensions (Fig. 7.3a) and the corresponding pattern by adding machining and shrinkage allowances (Fig. 7.3b) (see Examples 7.1–7.4).

Recently, Jiang and co-researchers (2024) have investigated the effect of geometric characteristics of castings on the linear contraction of sand casting

Table 7.1 Machining allowances for patterns for sand casting

Pattern dimension (mm)	Surface	Cope side	Bore
Cast iron			
Up to 150	2.4	4.8	3.2
151–300	3.2	6.4	3.2
301–500	4.0	6.4	4.8
501–900	4.8	6.4	6.4
901–1500	4.8	8.0	8.0
Cast steels			
Up to 150	3.2	6.4	3.2
151–300	4.8	6.4	6.4
301–500	6.4	8.0	6.4
501–900	6.4	9.6	7.0
901–1500	6.4	12.7	8.0
Nonferrous metals and alloys			
Up to 75	1.6	1.6	1.6
76–150	1.6	2.4	2.4
151–305	1.6	3.2	2.4
306–510	2.4	3.2	3.2
511–915	3.2	4.0	3.2
916–1520	3.2	4.8	4.0

Table 7.2 Linear shrinkages for some casting metals and alloys

Metal/alloy	Shrinkage (%)	Metal/alloy	Shrinkage (%)	Metal/alloy	Shrinkage (%)
Aluminum	1.0–1.5	Carbon steels	1.6–2.1	Magnesium	2.1
Gray cast iron	0.8–1.3	Stainless steels	2.1	Tin	2.1
White cast iron	2.1	Brass (yellow)	1.3–1.6	Zinc	2.6

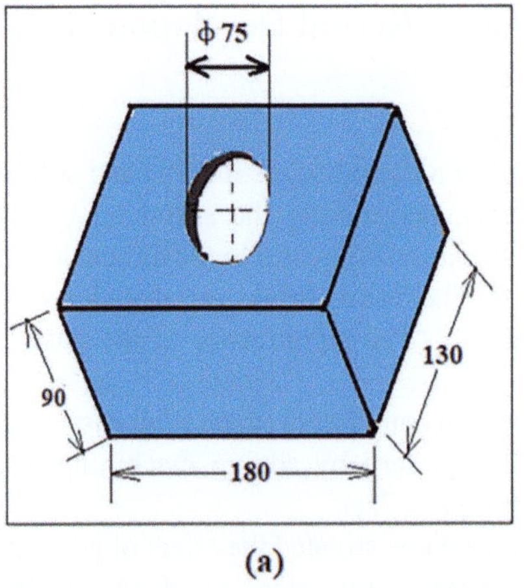

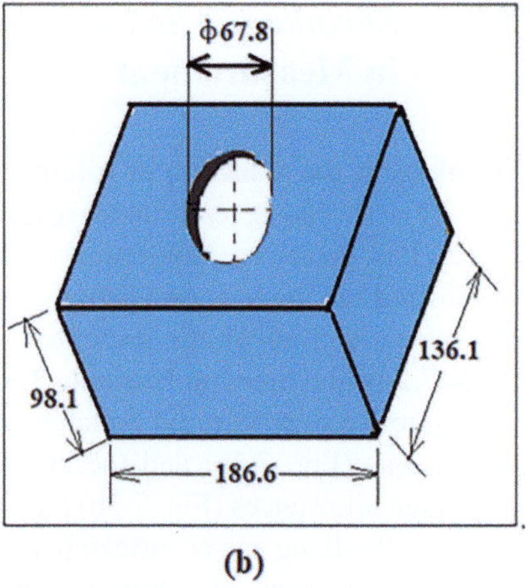

Fig. 7.3 The 3D sketches of a cast-iron (CI) component (**a**), and its pattern (**b**)

Fig. 7.4 Draft, draft angle, and radius in the pattern design

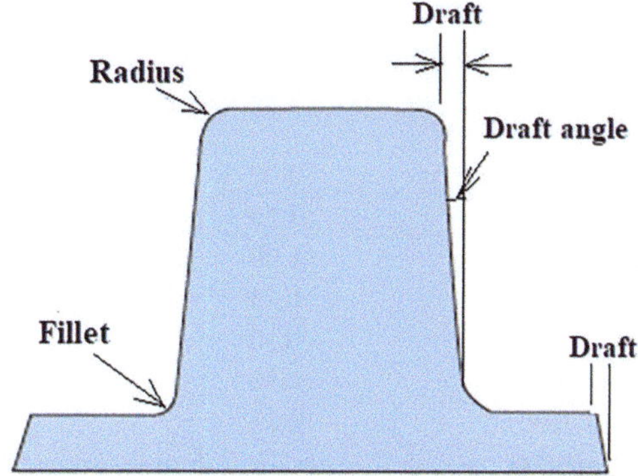

Mg–9Gd–3Y–0.5Zr alloy (VW93). They have reported that the dimensions of free and restrained structure castings can be precisely extracted by a 3D scanner and then used for the calculation of the contraction coefficient (Jiang et al., 2024).

7.2.4 Draft Allowance

The taper angle in the faces of a pattern, for its easy removal, is called the *draft allowance*. The vertical faces of the pattern are in continual contact with the sand during the withdrawal of the pattern from the sand mold. This feature may damage the mold cavity. It is therefore important to allow a draft allowance. Besides draft angles, the radii of the sharp edges must be properly examined in the pattern design (see Fig. 7.4). The draft angles for various height dimensions, for pattern design, are presented in Table 7.3.

7.3 Casting Gating System: *Recent Development*

A casting-gating system is the network of channels that allow molten metal to be poured into a mold and fill the mold cavity in a precisely-engineered manner to produce a sound and defect-free casting (Charbauski, 2024). The components of a gating system include: sprue (or down-sprue), choke, runner, and riser gates (see Fig. 7.5). Modern gating systems are fabricated by additive manufacturing (AM) technology. It has recently been reported that numerically optimized gating systems that can only be fabricated via 3D sand printing (3DSP) have the potential to signifi-cantly improve both mechanical and metallurgical performance of sand castings (Sama et al., 2019).

Table 7.3 Draft angles for pattern design

Height of given surface (mm)	Draft angle for external surface (deg)	Draft angle for internal surface (deg)
Wooden pattern		
Up to 20	3.00	3.00
21–50	1.50	2.50
51–100	1.00	1.50
101–200	0.75	1.00
201–300	0.50	1.00
301–800	0.50	1.75
801–2000	0.35	0.50
Over 2000	–	0.25
Plastic and metal		
Up to 20	1.50	3.00
21–50	1.00	2.00
51–100	0.75	1.00
101–200	0.50	0.75
201–300	0.50	0.75
301–800	0.35	0.50

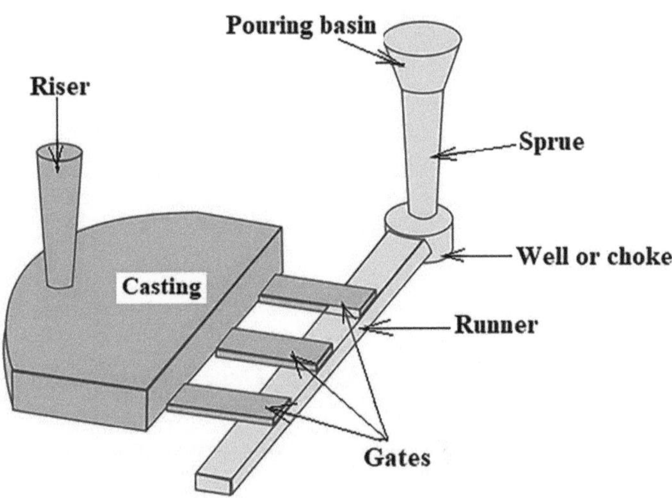

Fig. 7.5 A casting gating system

7.4   Sprue Design

7.4.1   Taper in Sprue Design: Recent Development

The sprue is the tool used to create a sprue hole to enable us to pour the molten metal into the sprue hole via a pouring basin (see Fig. 7.5). It was established in Chap. 2 that the use of a straight sprue in a casting gating system results in gas-aspiration defects (see Sect. 2.3.3). This problem can be overcome by using the law of continuity, i.e.:

$$Q = v_1 \cdot A_1 = v_2 \cdot A_2 \qquad (7.1)$$

where Q is the volume flow rate (m³/s), v_1 and v_2 are the flow velocities at the top and the bottom, respectively, and A_1 and A_2 are the cross-sectional areas at the top and bottom, respectively [see Fig. 7.6a, b]. The application of the *law of continuity*, for preventing air-aspiration casting defect, is explained in the following paragraph.

Consider the flow velocities v_1 and v_2. In the case of straight sprue (the cross-sectional areas are the same, i.e., $A_1 = A_2$), the flow velocity at point 2 is higher than that at point 1 ($v_2 > v_1$) (due to gravitational acceleration). As the velocity of the flowing metal increases toward the base of the sprue, air can be aspirated into the liquid and conducted into the mold cavity. Thus, the straight sprue (Fig. 7.6a) would result in air-aspiration casting defects. In order to overcome the air-aspiration problem, a smart solution has been proposed, as follows. According to this solution (smart sprue design), the taper (inclination) of the sprue will increase the friction between liquid metal and the mold, thereby decreasing the velocity at the bottom (base) of the sprue (Zarubin & Zarubina, 2017). In order to comply with the law of continuity (Eq. 7.1), the cross-sectional area A_2 must be proportionally smaller than A_1 ($A_2 < A_1$) (see Fig. 7.6b). Hence, a taper must be built into the sprue to ensure laminar flow of molten metal through the runner, thereby avoiding air-aspiration/ dross casting defects. Recently, Dojka and co-researchers (2022) conducted experiments on the optimization of down-sprue geometry in sand casting. They reported

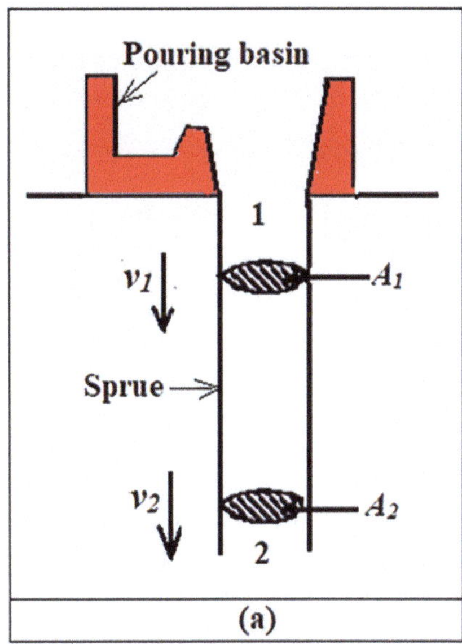

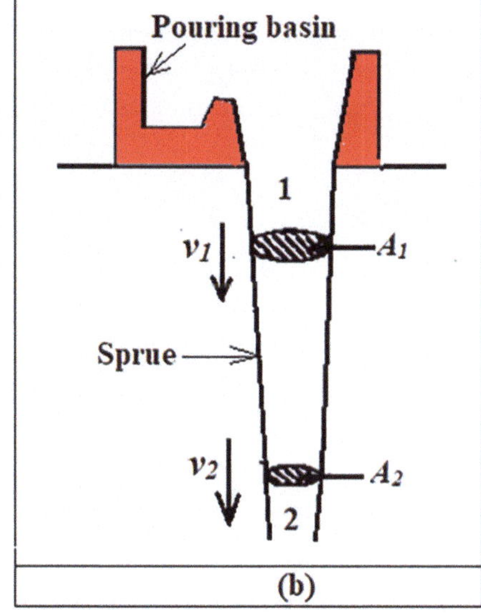

Fig. 7.6 Incorrect sprue design (**a**), and correct sprue design (**b**)

that short sprues can be manufactured with a straight-line taper, and long sprues will especially benefit from a hyperbolic taper (Dojka et al., 2022).

7.4.2 Sprue Design Rules

A good sprue design is important to prevent casting defects. The following rules should be followed to ensure good sprue design:

1. The sprue should be properly sized (length, top diameter, base diameters, etc.) to limit the flow rate of molten metal.
2. For large castings, rectangular cross-sectional sprues are better than circular ones.
3. For small- and medium-sized castings, circular sprues are preferred.
4. The sprue should be located as far from the gates as possible, since longer path enables the flow to become more laminar before it reaches the gate.
5. The provision of a well in the sprue design also helps to ensure laminar flow of molten metal (see Fig. 7.5). The cross-sectional area of the sprue well should be approximately five times that of the sprue exit, and its depth should be two times that of the runner's height.

7.4.3 Design Analysis

An expression for the velocity at the sprue base can be derived with reference to Fig. 7.7, as follows. Metal in the pouring basin (reservoir) can be assumed to be at zero velocity. According to the law of conservation of energy, the gain in kinetic energy is equal to the loss in potential energy, i.e.,

Fig. 7.7 Sprue design parameters

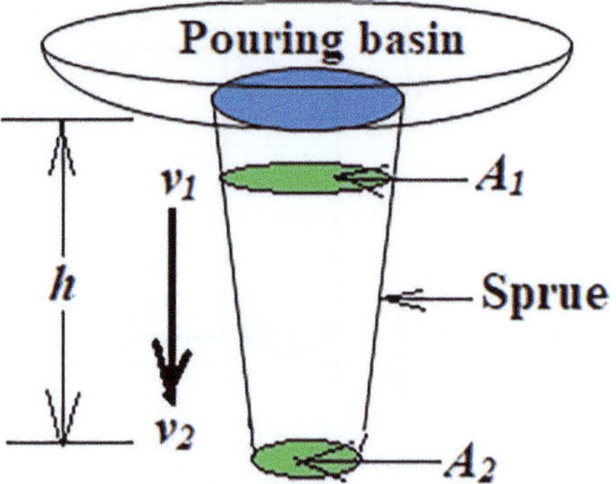

$$\frac{1}{2}m\left(v_2^2 - 0^2\right) = mgh \tag{7.2}$$

or

$$v_2 = \sqrt{2 \cdot g \cdot h} \tag{7.3}$$

where v_2 is the base velocity (m/s), h is the height of length of the sprue (m), and $g = 9.81$ m/s^2.

Once the base velocity (v_2) has been calculated by using Eq. 7.3, the velocity at the top of the sprue (v_1) can be determined by re-arranging the terms in Eq. 7.1, as follows:

$$v_1 = \frac{v_2 \cdot A_2}{A_1} = \frac{Q}{A_1} \tag{7.4}$$

The significance of Eqs. 7.1–7.4 is illustrated in Examples 7.5–7.9.

The time to fill the mold cavity can be calculated by:

$$T_{mf} = \frac{V}{Q} \tag{7.5}$$

where T_{mf} is the time to fill the mold cavity (s), V is the volume of casting (m^3), and Q is the volume flow rate (m^3/s) (see Example 7.10).

7.4.4   Reynolds Number for Sprue Design

Reynolds number (R_e) is a reliable quantitative technique for avoiding turbulent flow. It is numerically defined by the following formula:

$$R_e = \frac{v \cdot D \cdot \rho}{\eta} \tag{7.6}$$

where v is the velocity of the molten metal at the base of the sprue (m/s), D is the diameter of the channel (runner) (m), ρ is the density of the liquid metal (kg/m^3), and η is the viscosity of the liquid metal at the specified temperature (Pa s). The following rules apply to Eq. 7.6: (a) the flow is laminar (desirable) when $0 < R_e < 2000$, (b) the flow is transitional (harmless) when $2000 < R_e < 20{,}000$, and (c) the flow is turbulent (causing air entrainment defect) when $R_e > 20{,}000$ (see Example 7.11).

7.5   Riser Design

7.5.1   Riser Design Rules

A proper design of riser is important because an improperly designed riser would result in casting defects, particularly a shrinkage cavity (see Chap. 12). The following rules must be followed in designing a riser: (a) the metal in the riser must cool more slowly than the casting, and (b) the riser must have sufficient material to compensate for the casting shrinkage. According to the *Chvorinov's rule,* the slowest cooling (the greatest t_s) is achieved with the greatest volume (V) and the least surface area (A) (see Chap. 3, Eq. 3.11). This is why a cylindrical riser (instead of rectangular cross-section) is generally used in the casting design. Additionally, risers should be located near thick sections (high V/A ratio) of the casting (see Fig. 7.8). There are two types of riser: (a) top riser (see Fig. 7.1), and (b) side riser (see Fig. 7.2).

7.5.2   Engineering Analysis for Riser Design

The riser for a mold can be designed by using one of the following three methods: (a) Cain's methods, (b) modulus method, and (c) shape factor method. The modulus method is the simplest one. It is explained in the following paragraph.

Modulus is the ratio of volume to surface area, i.e., $M = V/A$. Accordingly, the modulus of casting (M_c) and the modulus of riser (M_r) can be mathematically expressed as:

$$M_c = \frac{V_c}{A_c} \tag{7.7a}$$

$$M_r = \frac{V_r}{A_r} \tag{7.7b}$$

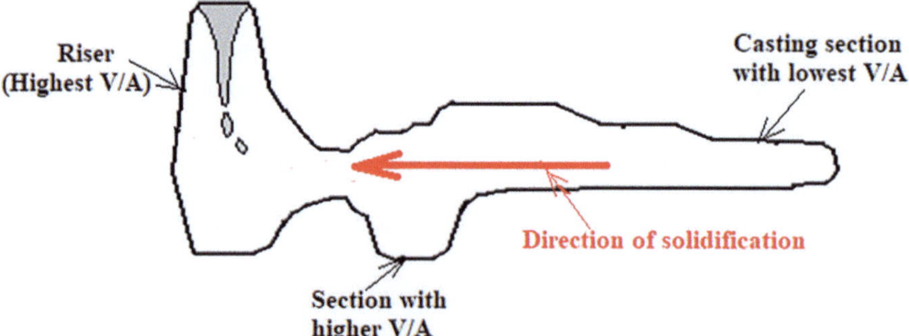

Fig. 7.8 Location of riser near the thick section of the casting

where V_c is the volume of casting, A_c is the surface area of the casting, V_r is the volume of riser, and A_r is the surface area of the riser. In order to produce a casting free of shrinkage defect, the riser must cool more slowly than the casting, which requires that the modulus of riser should be greater than the modulus of casting by a factor of 1.2, i.e.,

$$M_r = 1.2M_c \tag{7.8}$$

The modulus of cylinder is given by the following formula (see Fig. 7.9a):

$$M_{\text{cylinder}} = \frac{D \cdot H}{2 \times (D + 2H)} \tag{7.9}$$

For side risers, $H = D$; thus, Eq. 7.9 simplifies to (see Fig. 7.9b):

$$M_{D=H} = \frac{D}{6} \tag{7.10}$$

The significance of Eqs. 7.7a–7.10 is illustrated in Examples 7.12 and 7.13.

7.5.3   Recent Development in Riser Design

Although cylindrical risers are conventionally used in sand casting design, recent research has shown better results with different riser geometries. Recently, Shuvo and Manogharan (2021) have studied the effects of spherical and ellipsoid risers on casting yield. They have shown that the results for spherical riser performance showed a 7% increase in feeding time during solidification along with a 47.27%

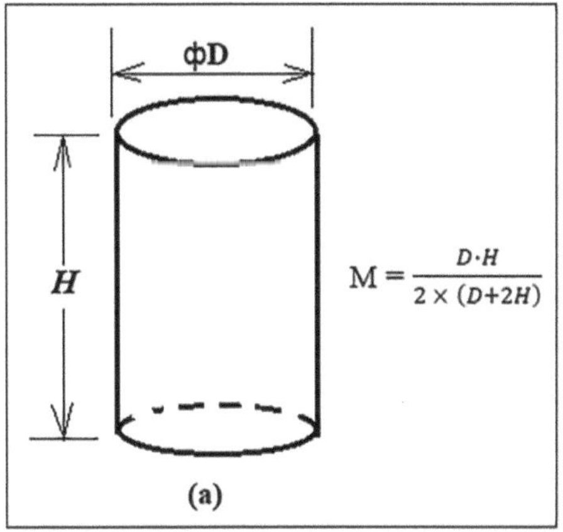

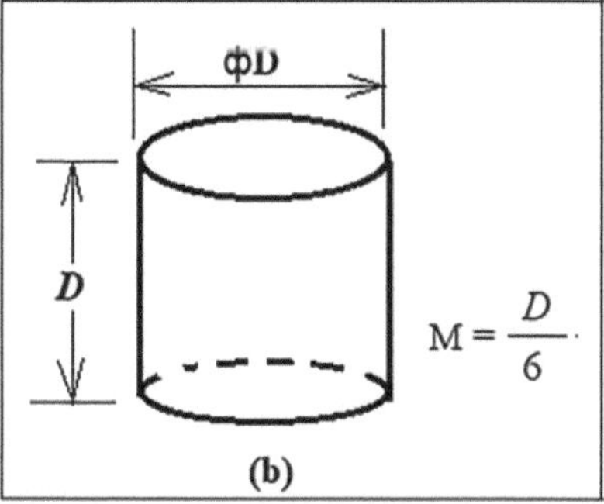

Fig. 7.9 Modulus of cylinder: (**a**) $H > D$, (**b**) $H = D$

reduction in entrained air volume fraction. The ellipsoid riser studied in this research also showed identical results with a 26.5% increase in casting yield (Shuvo & Manogharan, 2021).

7.6   Mold Design

7.6.1   Thrust on Cores in Sand Molds

A *core* is a piece of hardened sand used to create holes and various shaped cavities in a casting (see Fig. 7.2). When molten metal is poured, the buoyant force tends to displace the core, which causes core displacement resulting in casting defects. The buoyant force or thrust tends to lift the core vertically upward, while the weight of the core acts vertically downward. The net buoyant force can be determined by:

$$F_{\mathrm{b}} = W_{\mathrm{m}} - W_{\mathrm{c}} = \left(m_{\mathrm{m}} \cdot g \right) - \left(m_{\mathrm{c}} \cdot g \right) = \left[V \cdot \rho_{\mathrm{m}} - V \cdot \rho_{\mathrm{c}} \right] g = g \cdot V \cdot \left(\rho_{\mathrm{m}} - \rho_{\mathrm{c}} \right) \quad (7.11)$$

where F_{b} is the net buoyant force (N), W_{m} is the weight of the molten metal displaced by the core (N), W_{c} is the weight of the core (N), V is the volume of the core (= volume of metal displaced by the core) (m³), g is the gravitational acceleration ($g = 9.81$ m/s²), ρ_{m} is the density of metal (kg/m³), and ρ_{c} is the density of the core sand (kg/m³) (see Examples 7.14–7.16).

7.6.2   Chaplets to Support Cores

The buoyant force tends to lift the core, while the weight of the core is directed down. In order to prevent the displacement of the core, chaplets are used (see Fig. 7.2). Some chaplets are required above the core, and some chaplets are used beneath the core.

The number of chaplets required above the core (to resist buoyancy) can be found by (Huda, 2020):

$$n_{\mathrm{b}}^{\mathrm{ch}} = \frac{W_{\mathrm{m}}}{F_{\mathrm{ch}}} \quad (7.12)$$

where $n_{\mathrm{b}}^{\mathrm{ch}}$ is the number of chaplets required above the core to resist buoyancy, W_{m} is the weight of the metal displaced (N), and F_{ch} is the force sustained by one chaplet (N). The number of chaplets needed beneath the core (to resist the weight of the core) can be calculated by:

$$n_{\mathrm{w}}^{\mathrm{ch}} = \frac{W_{\mathrm{c}}}{F_{\mathrm{ch}}} \quad (7.13)$$

where n_w^{ch} is the number of chaplets required beneath the core to resist the weight of core, and W_c is the weight of the core (N) (see Example 7.17).

7.6.3   Metallostatic Force in Sand Molds

The molten metal, in a sand mold, can exert considerable lifting force on a cope (see Fig. 7.10). This lifting force is called the *metallostatic force*, which can be determined by:

$$F_m = A_p \cdot g \cdot \rho_m \cdot H \qquad (7.14)$$

where F_m is the metallostatic force (N), A_p is the projection area of the casting in the parting plane (m²), ρ_m is the density of the metal (kg/m³), and H is the metal head ($H = h - c$) (in m).

7.6.4   Calculating Weight Required to Prevent Lifting of Cope

The metallostatic force F_m tends to lift the cope, which may cause casting defects and safety issues to the workers. It is therefore important to prevent lifting of the cope by placing pre-calculated weight on the top surface of the cope. Besides F_m, there is a gravity force, i.e., weight of sand in the cope, which acts downwards. The weight of sand in the cope can be calculated by:

$$W_s = V_s \cdot \rho_s \cdot g \qquad (7.15)$$

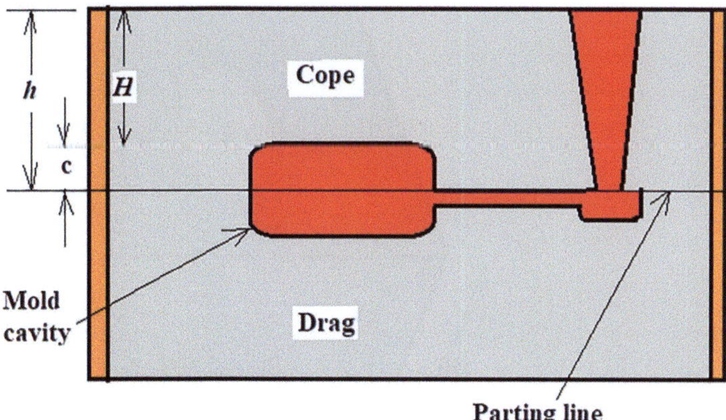

Fig. 7.10 The metal head H in a sand mold (h = height of the sprue, c = height of the casting in the cope)

where W_s is the weight of sand in the cope (N), V_s is the volume of sand in the cope (m³), ρ_s is the density of sand (kg/m³), and $g = 9.81$ m/s². The weight required to prevent lifting of cope can be calculated by:

$$W = F_m - W_s \tag{7.16}$$

where W is the weight required to prevent lifting of cope (N). The mass m (in kgf) needed to be placed on the top surface of the cope can be determined by:

$$m = \frac{W}{g} \tag{7.17}$$

The significance of Eqs. 7.14–7.17 is illustrated in Examples 7.18–7.25.

7.7   Worked Numerical Examples in Casting Design

Example 7.1 Determining the Machining Allowance for a Casting Drawing
Figure 7.11 shows the 3D sketch (with dimensions) of a gray-cast-iron component with a bore, which is required to be manufactured by sand casting. Determine the machining allowance for each dimension in Fig. 7.11.

Solution
The front view of Fig. 7.11 is shown as the 2D drawing in Fig. 7.12.

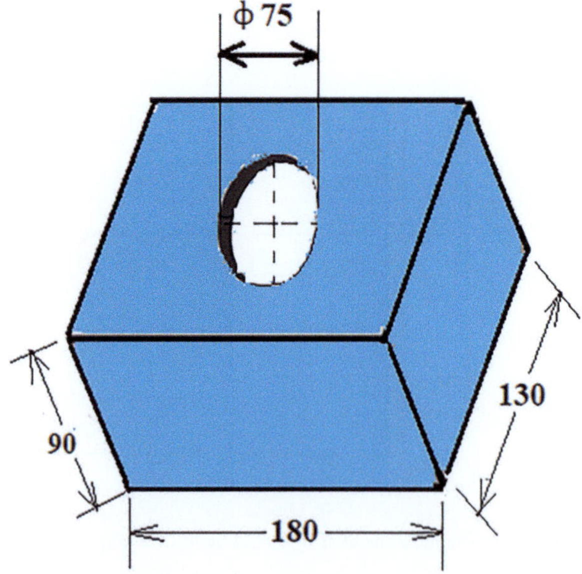

Fig. 7.11 The drawing of the cast-iron casting for Example 7.1 (dimensions in mm)

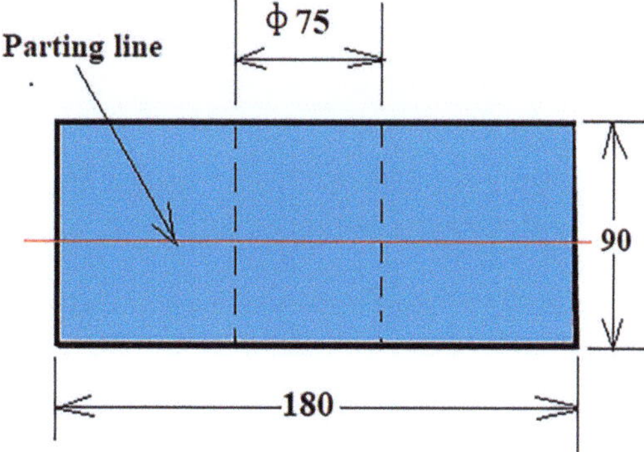

Fig. 7.12 The front view of the casting shown in Fig. 7.11 (dimensions in mm)

Since all dimensions are less than 150 mm, the reference to the cast-iron data in Table 7.1 yields:

Machining allowance for surface = 2.4 mm.
Machining allowance for cope side = 4.8 mm.
Machining allowance for bore = 3.2.

The machining allowances for each dimension are calculated as follows:

For the $90 - mm$ dimension, the machining allowance $= 2.4 + 4.8 = 7.2\,\text{mm}$

For the 180-mm dimension, the machining allowance $= 2.4\ \text{mm}$

For the 130-mm dimension, the machining allowance $= 2.4\,\text{mm}$

For the 75-mm bore dia., the machining allowance $= -(2 \times 3.2) = -6.4\,\text{mm}$

Example 7.2 Determining the New Dimensions by Adding the Machining Allowances

By using the data in Example 7.1, determine the new dimensions by adding machining allowance to each original dimension.

Solution

For the 90-mm dimension, the new dimension $= 90 + 7.2 = 97.2\,\text{mm}$

For the 180-mm dimension, the new dimension $= 180 + 2.4 = 182.4\,\text{mm}$

For the 130-mm dimension, the new dimension $= 130 + 2.4 = 132.4\,\text{mm}$

For the 75-mm bore dia., the new dimension $= 75 - (2 \times 3.2) = 68.6\,\text{mm}$

Example 7.3 Determining the Shrinkage Allowances for Each Dimension
By using the data for the new dimensions in Example 7.2, determine the shrinkage allowances for each dimension.

Solution
By reference to Table 7.2, the shrinkage for gray cast iron may be taken as 1.0%.
The shrinkage allowances, for each new dimension, are calculated as follows:

$$\text{For the } 97.2\text{-mm dimension, the shrinkage allowance} = 1.0\% \times 97.2 = 0.972\,\text{mm}$$

$$\text{For the } 182.4\text{-mm dimension, the shrinkage allowance} = 1.0\% \times 182.4 = 1.824\,\text{mm}$$

$$\text{For the } 132.4\text{-mm dimension, the shrinkage allowance} = 1.0\% \times 132.4 = 1.324\,\text{mm}$$

$$\text{For the } 68.6\text{-mm bore dia., the shrinkage allowance} = -1.0\% \times 68.6 = -0.67\,\text{mm}$$

Example 7.4 Drawing the Pattern Sketch (with Dimensions) Based on Pattern Allowances
By using the data in Examples 7.1–7.3, determine the final pattern dimensions (in mm) taking into consideration both machining allowance and shrinkage allowance, and hence draw the 3D sketch of the final pattern showing all dimensions.

Solution
By using the data in Examples 7.1 and 7.2, we can construct the new table, as Table 7.4.
The sketch of the pattern with all dimensions is shown in Fig. 7.13.

Example 7.5 Determining the Velocity of Molten Metal at the Sprue Bottom
The height of a sprue is 10 cm, and its cross-sectional areas at the top and the bottom are 4.2 and 2.3 cm², respectively. The cross-sectional area of the pouring basin is 40 cm². Determine the velocity of molten metal at the bottom of the sprue.

Solution
$h = 10$ cm, $A_1 = 4.2$ cm², $A_2 = 2.3$ cm², $A_{\text{basin}} = 40$ cm², $g = 981$ cm/s², $v_2 = ?$
By using Eq. 7.3,

$$v_2 = \sqrt{2 \cdot g \cdot h} = \sqrt{2 \times 981 \times 10} = \sqrt{19,620} = 140.1\,\text{cm/s} = 1.4\,\text{m/s}$$

Table 7.4 Pattern dimensions (in mm) calculation based on Example 7.3

Original dimension	Machining allowance	Shrinkage allowance	Pattern dimension
90	7.2	0.97	98.17
180	2.4	1.82	184.22
130	2.4	1.32	133.72
75 bore	−6.4	−0.67	67.93 ≅ 68

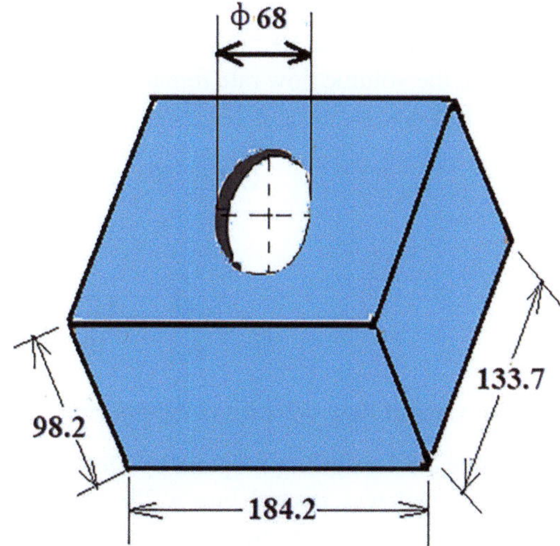

Fig. 7.13 The sketch of the pattern with final dimensions (in mm) for Example 7.4

The velocity of molten metal at the bottom of the sprue = v_2 = 1.4 m/s.

Example 7.6 Determining the Volume Flow Rate of the Molten Metal
By using the data in Example 7.5, determine the volume flow rate of the molten metal.

Solution
A_1 = 4.2 cm^2, A_2 = 2.3 cm^2 = 2.3 × 10^{-4} m^2, v_2 = 1.4 m/s, v_1 =?
   By using Eq. 7.1,

$$Q = v_2 \cdot A_2 = 1.4 \times 2.3 \times 10^{-4} = 3.22 \times 10^{-4}\, \text{m}^3/\text{s}$$

The volume flow rate of molten metal = Q = 3.22 × 10^{-4} m^3/s.

Example 7.7 Determining the Velocity of Molten Metal at the Sprue Top
By using the data in Examples 7.5 and 7.6, determine the velocity of metal at the top of the sprue.

Solution
Q = 3.22 × 10^{-4} m^3/s, A_1 = 4.2 cm^2 = 4.2 × 10^{-4} m^2, v_1 =?
   By using Eq. 7.4,

$$v_1 = \frac{Q}{A_1} = \frac{3.22 \times 10^{-4}}{4.2 \times 10^{-4}} = 0.767\, \text{m}/\text{s}$$

Example 7.8 Determining the Vertical Flow Velocity of Metal Within the Pouring Basin
By using the data in Examples 7.5 and 7.7, determine the vertical flow velocity of the molten metal within the pouring basin.

Solution

$A_{\text{basin}} = 40$ cm$^2 = 40 \times 10^{-4}$ m^2, $Q = 3.22 \times 10^{-4}$ m^3/s, $v_{\text{basin}} = ?$

Since the volume flow rate remains constant (see Fig. 7.6), Eq. 7.1 can be modified as follows:

$$Q = v_1 \cdot A_1 = v_2 \cdot A_2 = v_{\text{basin}} \cdot A_{\text{basin}}$$

or

$$v_{\text{basin}} = \frac{Q}{A_{\text{basin}}} = \frac{3.22 \times 10^{-4}}{40 \times 10^{-4}} = 0.08 \, \text{m/s}$$

The vertical flow velocity in the pouring basin = 0.08 m/s.

Example 7.9 Designing a Sprue by Determining Its Base Diameter

Molten metal was poured into a 25-cm-long sprue at a volumetric flow rate of 45 cm^3/s. Design the sprue by determining its base diameter in order to keep the same flow rate.

Solution

$h = 25$ cm, $Q = 45$ cm^3/s, $A_2 = ?$, $D = ?$

By using Eq. 7.3,

$$v_2 = \sqrt{2 \cdot g \cdot h} = \sqrt{2 \times 981 \times 25} = \sqrt{49,050} = 221.47 \, \text{cm/s}$$

By using the modified form of Eq. 7.1,

$$A_2 = \frac{Q}{v_2} = \frac{45}{221.47} = 0.203 \, \text{cm}^2$$

or

$$\frac{\pi}{4} D^2 = A_2 = 0.203$$

$$D^2 = \frac{4}{\pi} \times 0.203 = 0.2585$$

$$D = \sqrt{0.2585} = 0.51 \, \text{cm} = 5.1 \, \text{mm}$$

The base diameter of the sprue = 5.1 mm.

Example 7.10 Determining the Time to Fill the Mold Cavity

By using the data in Example 7.8, determine the time to fill the mold cavity. The volume of casting is 0.008 m^3.

Solution

$V = 0.008$ m^3, $Q = 3.22 \times 10^{-4}$ m^3/s, $T_{\text{mf}} = ?$

By using Eq. 7.5,

$$T_{\mathrm{mf}} = \frac{V}{Q} = \frac{0.008}{3.22 \times 10^{-4}} = 24.8\,\mathrm{s}$$

The time to fill the mold cavity $= T_{\mathrm{mf}} = 24.8$ s, which is reasonably short (good).

Example 7.11 Predicting the Type of Flow Based on Reynolds Number for Sand Casting

An 8-cm-long sprue is designed for casting a metallic component. The density of the metal is 2.8 g/cm³. The viscosity of the liquid metal is 10^{-3} Pa s. The channel diameter is 0.3 cm. Predict the type of flow (laminar, transition, turbulent). Will this sprue design cause air-entrainment/dross defect?

Solution

$h = 8$ cm $= 0.08$ m, $D = 0.3$ cm $= 0.003$ m, $\rho = 2.8 \times 10^3$ kg/m³, $\eta = 10^{-3}$ Pa s $= 0.001$ Pa s, $R_e = ?$

$$v = v_2 = \sqrt{2 \cdot g \cdot h} = \sqrt{2 \times 9.81 \times 0.08} = 1.253\,\mathrm{m/s}$$

By using Eq. 7.6,

$$R_e = \frac{v \cdot D \cdot \rho}{\eta} = \frac{1.253 \times 0.003 \times 2.8 \times 10^3}{0.001} = 10{,}525.2$$

Since R_e value is in the range of 2000–20,000, the flow is transitional. This sprue design will not cause air-entrainment/dross casting defect.

Example 7.12 Designing a Side Riser for Sand Casting

Design a side riser for a sand mold for the casting with dimensions: 30 cm × 25 cm × 15 cm. Take the height of the riser equal to its diameter.

Solution

The sketch of the casting with dimensions is shown in Fig. 7.14.

$$V_c = 30\,\mathrm{cm} \times 25\,\mathrm{cm} \times 15\,\mathrm{cm} = 11{,}250\,\mathrm{cm}^3$$

Fig. 7.14 The casting dimensions for Example 7.12

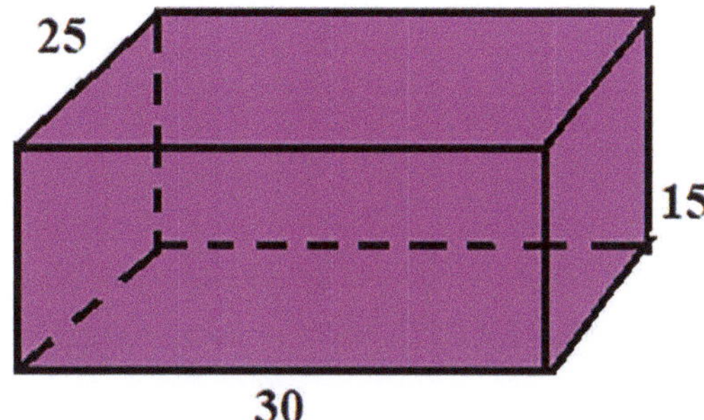

$$\text{Surface area of casting} = A_{\mathrm{c}} = \left[2\times(30\times25)\right] + \left[2\times(25\times15)\right]$$
$$+ \left[2\times(30\times15)\right] = 3150\,\mathrm{cm}^2$$

$$M_{\mathrm{c}} = \frac{V_{\mathrm{c}}}{A_{\mathrm{c}}} = \frac{11{,}250}{3150} = 3.57$$

By using Eq. 7.8,

$$M_{\mathrm{r}} = 1.2 M_{\mathrm{c}} = 1.2 \times 3.57 = 4.286$$

Since $H = D$ for the side riser, Eq. 7.10 is applicable, i.e.:

$$M_{\mathrm{r}} = 4.286 = \frac{D}{6}$$

$$D = 4.286 \times 6 = 25.7\,\mathrm{cm}$$

$$H = D = 25.7\,\mathrm{cm}$$

The riser design specifications: diameter = 25.7 cm, height = 25.7 cm.

Example 7.13 Designing a Top Riser for Sand Casting
Repeat Example 7.12 for a top riser with its height doubled its diameter.

Solution
$M_{\mathrm{r}} = 4.286$, $H = 2D$, $D =?$

By using Eq. 7.9,

$$M_{\text{cylinder}} = \frac{D \cdot H}{2\times(D+2H)} = M_{\mathrm{r}} = 4.286$$

$$\frac{D \cdot (2D)}{2\times(D+2\times2D)} = 4.286$$

$$\frac{2D^2}{10D} = \frac{D^2}{5D} = 4.286$$

$$D^2 = (5D) \times 4.286 = 21.43 D$$

$$D = 21.43\,\mathrm{cm}$$

$$H = 2D = 2 \times 21.43 = 42.86\,\mathrm{cm}$$

The riser design specifications: diameter = 21.43 cm, height = 42.86 cm.

Example 7.14 Determining the Buoyant Force When the Volume of Core Is Known

The volume of a sand core, located inside a mold cavity, is 2000 cm³. The core is used in the sand casting of a cast-iron component. Determine the net buoyant force acting on the core. The density of cast iron is 7.16 g/cm³, and the sand density is 1.6 g/cm³.

Solution

$V = 2000$ cm³ $= 2000 \times 10^{-6}$ m³, $\rho_m = 7.16$ g/cm³ $= 7.16 \times 10^3$ kg/m³, $\rho_c = 1.6$ g/cm³ $= 1.6 \times 10^3$ kg/m³, $F_b =?$

By using Eq. 7.11

$$F_b = g \cdot V \cdot (\rho_m - \rho_c) = 9.81 \times 2000 \times 10^{-6} \times (7.16 \times 10^3 - 1.6 \times 10^3) = 109.1\,\text{N}$$

The net buoyant force $= F_b = 110$ N.

Example 7.15 Determining the Buoyant Force When the Volume of Core Is Unknown

A copper casting is manufactured in a sand mold using a sand core weighing 25 kgf. Determine the net buoyant force acting on the core during pouring. The density of liquid copper is 8.02 g/cm³.

Solution

$m_c = 25$ kgf, $\rho_c = 1.6$ g/cm³ $= 1.6 \times 10^3$ kg/m³, $\rho_m = 8.02 \times 10^3$ kg/m³, $F_b =?$

The density is defined as the ratio of mass to volume. Accordingly,

$$\text{Volume of core} = V_c = \frac{m_c}{\rho_c} = \frac{25}{1.6 \times 10^3} = 0.01562\,\text{m}^3$$

By using Eq. 7.11

$$F_b = g \cdot V \cdot (\rho_m - \rho_c) = 9.81 \times 0.01562 \times (8.02 \times 10^3 - 1.6 \times 10^3) = 983.75\,\text{N}$$

Example 7.16 Determining the Mass of a Casting When the Thrust on a Core Is Known

The net buoyant force on a sand core, in a mold cavity, is 200 N. The volume of the mold cavity forming the outside surface of the casting is 4700 cm³. Determine the mass of the steel casting.

The densities of the core sand and steel are 1.6 and 7.8 g/cm³, respectively.

Solution

$F_b = 200$ N, $\rho_c = 1600$ kg/m³, $\rho_m = 7800$ kg/m³, $m_m =?$

Volume of outside surface of the mold cavity $= 4700$ cm³ $= 4700 \times 10^{-6}$ m³ $= 0.0047$ m³.

By using the modified form of Eq. 7.11,

$$V = \frac{F_b}{g \cdot (\rho_m - \rho_c)} = \frac{200}{9.81 \times (7800 - 1600)} = \frac{200}{9.81 \times 6200} = 0.003288 \, \text{m}^3$$

Volume of core = 0.003288 m³

Volume of casting = Volume of outside surface of mold cavity − Volume of core

$$\text{Volume of casting} = 0.004700 - 0.003288 = 0.001412 \, \text{m}^3 = V_m$$

$$\text{Mass of steel casting} = m_m = \rho_m \cdot V_m = 7800 \times 0.001412 = 11 \, \text{kg}$$

Example 7.17 Determining the Number of Chaplets Needed to Support a Core

A brass component, with a hole is to be sand cast. The volume of core is 4800 cm³. The design of chaplets allows each chaplet to sustain a force of 40 N. The mold design must avoid core displacement during pouring of molten brass. Determine the number of chaplets that are needed: (a) above the core, and (b) beneath the core. The densities of brass and sand are 8.62 and 1.65 g/cm³, respectively.

Solution

F_{ch} = 40 N, V_c = 4800 cm³ = 4800 × 10⁻⁶ m³, ρ_m = 8620 kg/m³, and ρ_c = 1650 kg/m³

$$\text{Mass of core} = m_c = \rho_c \cdot V_c = 1650 \times 4800 \times 10^{-6} = 7.92 \, \text{kg}$$

$$W_c = m_c g = 7.92 \times 9.81 = 77.7 \, \text{N}$$

The volume of metal displaced is equal to the volume of core, i.e., $V_m = V_c = 4800 \times 10^{-6}$ m³

$$\text{Mass of metal} = m_m = \rho_m \cdot V_m = 8620 \times 4800 \times 10^{-6} = 41.37 \, \text{kg}$$

$$W_m = m_m \, g = 41.37 \times 9.81 = 405.84 \, \text{N}$$

(a) By using Eq. 7.12,

$$n_b^{ch} = \frac{W_m}{F_{ch}} = \frac{405.84}{40} = 10.15$$

The number of chaplets needed above the core = 11.

(b) By using Eq. 7.13,

$$n_w^{ch} = \frac{W_c}{F_{ch}} = \frac{77.7}{40} = 1.94$$

The number of chaplets needed beneath the core = 2.

Example 7.18 Determining the Metallostatic Force for a Split Pattern in Sand Mold

The split-pattern dimensions for a steel casting are shown in Fig. 7.15. The molten metal is poured, in the sand mold, at a steady flow rate of 45 cm³/s. The cope dimensions are 300 mm × 300 mm × 220 mm. The density of liquid steel is 7040 kg/m³, and its viscosity is 0.01 Pa·s. The sand density is 1600 kg/m³. Determine the metallostatic force acting on the cope.

Solution

The two-piece (split) pattern placed in the sand mold is schematically shown in Fig. 7.16.

h = sprue height = 220 mm, c = 40 mm, $H = h - c = 220 - 40 = 180$ mm = 0.18 m (see Fig. 7.16), g = 9.81 m/s², A_p = 160 mm × 160 mm = 0.16 m × 0.16 m = 0.0256 m², ρ_m = 7040 kg/m³, F_m =?

By using Eq. 7.14,

$$F_m = A_p \cdot g \cdot \rho_m \cdot H = 0.0256 \times 9.81 \times 7040 \times 0.18 = 318\,\text{N}$$

The metallostatic force = F_m = 318 N.

Example 7.19 Determining the Weight of Sand in the Cope of a Sand Mold for a Split Pattern

By using the data in Example 7.18, determine the weight of sand in the cope.

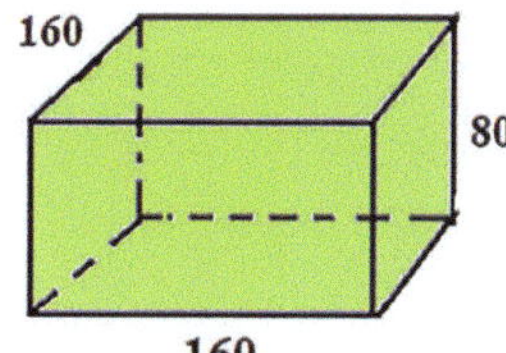

Fig. 7.15 The pattern dimensions for Example 7.18

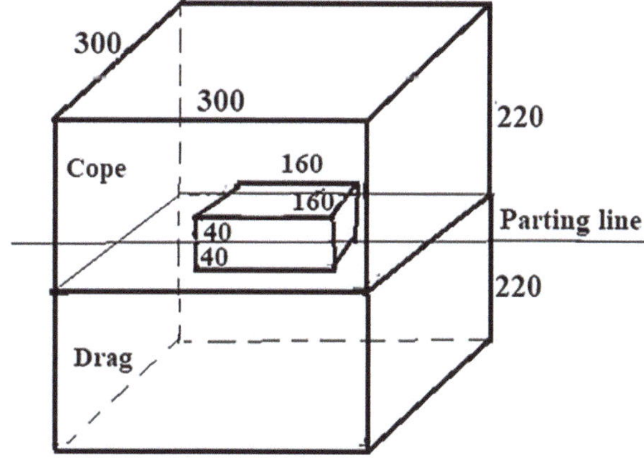

Fig. 7.16 The split (two-piece) pattern placed in sand mold for Example 7.18

Solution

Density of sand $= \rho_s = 1600$ kg/m^3. By reference to Fig. 7.16,

V_s = Volume of sand in the cope = Volume of cope − volume of pattern in the cope

$$V_s = (300 \times 300 \times 220) - (160 \times 160 \times 40)$$
$$= (0.30\,\text{m} \times 0.30\,\text{m} \times 0.22\,\text{m}) - (0.16 \times 0.16 \times 0.04)$$

$$V_s = 0.0198\,\text{m}^3 - 0.001024\,\text{m}^3 = 0.018776\,\text{m}^3$$

By using Eq. 7.15,

$$W_s = V_s \cdot \rho_s \cdot g = 0.018776 \times 1600 \times 9.81 = 294.7\,\text{N}$$

The weight of sand in the cope $= W_s = 294.7$ N.

Example 7.20 Determining the Weight Required to Be Placed on the Cope
By using the data in Examples 7.18 and 7.19, determine the weight (kgf) needed to be placed on the top surface of the cope to prevent its lifting.

Solution
$W_s = 294.7$ N, $F_m = 318$ N, $m =?$
By using Eq. 7.16,

$$W = F_m - W_s = 318 - 294.7 = 23.3\,\text{N}$$

By using Eq. 7.17,

$$m = \frac{W}{g} = \frac{23.3}{9.81} = 2.37\,\text{kg}$$

The weight needed to be placed on the top surface of the cope to prevent its lifting = 3 kgf.

Example 7.21 Determining the Metallostatic Force for a One-Piece Pattern in Sand Mold
Repeat Example 7.18 with the use of a one-piece pattern of the same dimensions.

Solution
The one-piece pattern placed in the sand mold is schematically shown in Fig. 7.17.

h = sprue height = 220 mm, $c = 0$, $H = h - c = 220 - 0 = 220$ mm = 0.22 m (see Fig. 7.17).

$g = 9.81$ m/s^2, $A_p = 160$ mm $\times$ 160 mm = 0.16 m $\times$ 0.16 m = 0.0256 m^2, $\rho_m = 7040$ kg/m^3, $F_m =?$
By using Eq. 7.14,

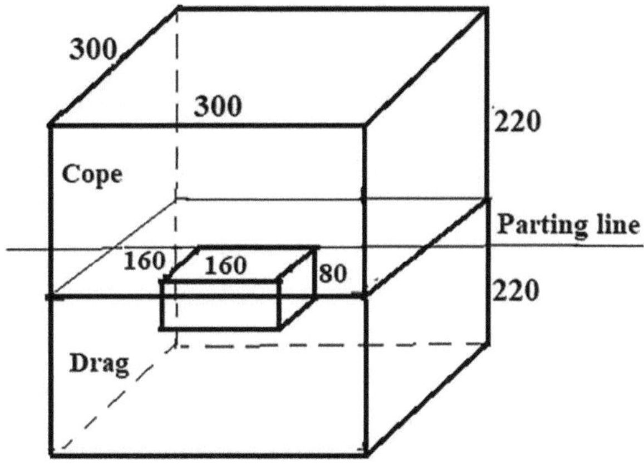

Fig. 7.17 One-piece pattern in sand mold for Example 7.21

$$F_\mathrm{m} = A_\mathrm{p} \cdot g \cdot \rho_\mathrm{m} \cdot H = 0.0256 \times 9.81 \times 7040 \times 0.22 = 389\,\mathrm{N}$$

The metallostatic force $= F_\mathrm{m} = 389$ N.

Example 7.22 Determining the Weight of Sand in the Cope for a One-Piece Pattern

By using the data in Example 7.21, determine the weight (in kgf) required to be placed on the top of the cope to prevent its lifting.

Solution

$F_\mathrm{m} = 389$ N, $\rho_\mathrm{s} = 1600$ kg/m^3, $W =$?

By reference to Fig. 7.17,

V_s = Volume of sand in the cope = Volume of cope – volume of pattern in the cope

$$V_\mathrm{s} = (300 \times 300 \times 220) - 0 = 0.30\,\mathrm{m} \times 0.30\,\mathrm{m} \times 0.22\,\mathrm{m} = 0.0198\,\mathrm{m}^3$$

By using Eq. 7.15,

$$W_\mathrm{s} = V_\mathrm{s} \cdot \rho_\mathrm{s} \cdot g = 0.0198 \times 1600 \times 9.81 = 310.78\,\mathrm{N}$$

By using Eq. 7.16,

$$W = F_\mathrm{m} - W_\mathrm{s} = 389 - 310.78 = 78.22\,\mathrm{N}$$

By using Eq. 7.17,

$$m = \frac{W}{g} = \frac{78.22}{9.81} = 8\,\mathrm{kgf}$$

The weight needed to be placed on the top surface of the cope to prevent its lifting = 8 kgf.

Example 7.23 Determining the Sprue Base Diameter When the Cope Dimensions Are Given

By using the data in Example 7.18, determine the sprue base diameter for the casting design.

Solution

By reference to Fig. 7.16, h = 220 mm = 22 cm. Q = 45 cm^3/s, g = 981 cm/s^2, A_2 =?, D =?

$$v_2 = \sqrt{2gh} = \sqrt{2 \times 981 \times 22} = 207.76\,\text{cm/s}$$

$$A_2 = \frac{Q}{v_2} = \frac{45}{207.76} = 216.6\,\text{cm}^2$$

$$\frac{\pi}{4} = D^2 = A_2 = 216.6$$

$$D^2 = 275.7\,\text{cm}^2$$

$$D = \sqrt{275.7} = 16.6\,\text{cm}$$

The sprue base diameter = D = 16.6 cm.

Example 7.24 Determining the Metallostatic Force for a Complex Component (Casting)

The pattern dimensions for a brass component are shown in Fig. 7.18. The component is required to be made by sand casting. Assume that the parting line is located 25 mm above the base of the pattern. Each flask dimensions are 350 mm × 350 mm

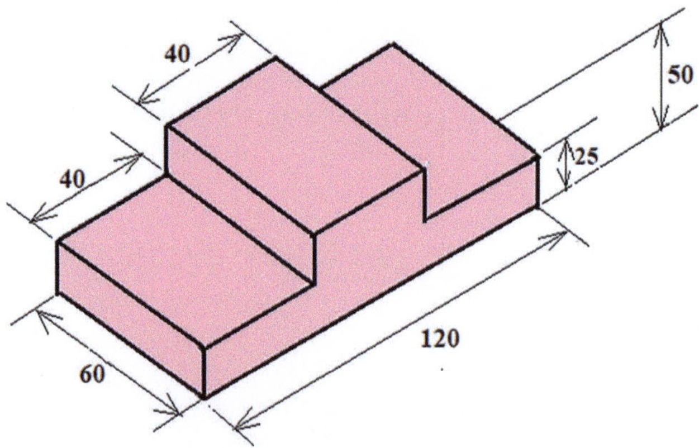

Fig. 7.18 Pattern dimensions for Example 7.24

× 180 mm. The density of brass is 8.6 g/cm^3, and the density of sand is 1.65 g/cm^3. Determine the metallostatic force.

Solution
Figure 7.19a, b shows the 3D sketch of the pattern (Fig. 7.19a) and 2D sketch of pattern's placement in the sand mold along with flask dimensions (see Fig. 7.19b).

h = sprue height = 180 mm, c = 25 mm, $H = h - c = 180 - 25 = 155$ mm = 0.15 5 m, g = 9.81 m/s^2, A_p = 120 mm × 60 mm = 0.12 m × 0.06 m = 0.0072 m^2, ρ_m = 8600 kg/m^3, F_m =?

By using Eq. 7.14,

$$F_m = A_p \cdot g \cdot \rho_m \cdot H = 0.0072 \times 9.81 \times 8600 \times 0.155 = 94.1\,\text{N}$$

The metallostatic force = F_m = 94.1 N.

Example 7.25 Determining the Weight Required to Be Placed on Cope for Complex Pattern
By using the data in Example 7.24, determine the weight (in kgf) required to be placed on the top of the cope to prevent its lifting.

Solution
F_m = 94.1 N, ρ_s = 1650 kg/m^3, W =?

By reference to Fig. 7.19,

V_s = Volume of sand in the cope = Volume of cope – volume of pattern in the cope

$$V_s = \left(350 \times 350 \times 180\right) - \left(60 \times 40 \times 25\right)$$

$$V_s = \left(0.35\,\text{m} \times 0.35\,\text{m} \times 0.18\,\text{m}\right) - \left(0.06 \times 0.04 \times 0.025\right)$$
$$= 0.02205\,\text{m}^3 - 0.00006\,\text{m}^3 = 0.02199\,\text{m}^3$$

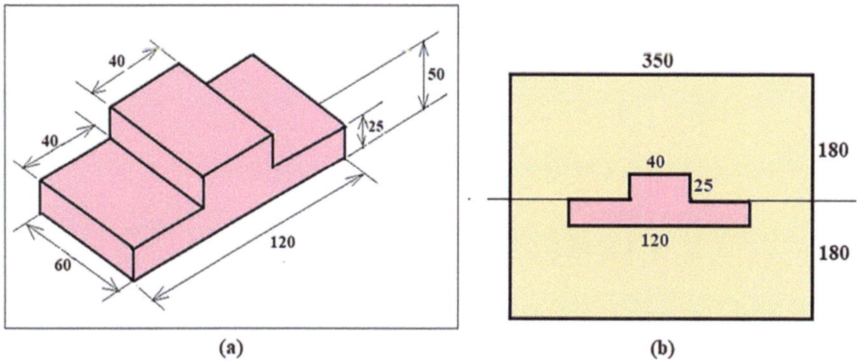

Fig. 7.19 The 3D sketch of pattern (**a**) and its placement in the sand mold (**b**)

By using Eq. 7.15,

$$W_s = V_s \cdot \rho_s \cdot g = 0.02199 \times 1650 \times 9.81 = 356\,\text{N}$$

By using Eq. 7.16,

$$W = F_m - W_s = 94.1 - 356 = -262\,\text{N}$$

The required weight is negative. It means that no weight is required to be placed on the cope.

Example 7.26 Designing a Sprue to Avoid Air Aspiration Defect in the Casting
The cross-sectional area at the top of the sprue is 900 mm^2, and its length is 180 mm. The flow rate of liquid metal into the sprue of a mold is 1 L/s. Design the sprue to avoid aspiration of the molten metal.

Solution
$A_1 = 900$ mm^2, $h = 180$ mm, $Q = 1$ L/s $= 1{,}000{,}000$ mm^3/s, $g = 9810$ mm/s^2

$$\text{Base velocity} = v_2 = \sqrt{2gh} = \sqrt{2 \times 9810 \times 180} = 1879.2\,\text{mm}/\text{s}$$

In order to avoid the air aspiration defect, the flow rate at the top and the bottom must be equal.

$$Q = A_1 * v_1 = A_2 * v_2$$

$$1{,}000{,}000 = A_2 * v_2 = A_2 \times 1879.2$$

$$A_2 = \frac{1{,}000{,}000}{1879.2} = 532.14\,\text{mm}^2$$

or

$$\frac{\pi}{4}D^2 = A_2 = 532.14$$

$$D^2 = \frac{4}{\pi} \times 532.14 = 677.54$$

$$D = \sqrt{677.54} = 26\,\text{mm} = 2.6\,\text{cm} = \text{Base diameter}$$

$$\text{Top cross sectional area} = A_1 = 900\,\text{mm}^2$$

$$\frac{\pi}{4}D^2 = A_1 = 900$$

$$D^2 = \frac{4}{\pi} \times 900 = 1146$$

$$D = \sqrt{1146} = 33\,\text{mm} = 3.3\,\text{cm} = \text{Top diameter}$$

The final design specifications for the sprue:

$$\text{Length} = 18\,\text{cm}, \text{Top diameter} = 3.3\,\text{cm}, \text{Base diameter} = 2.6\,\text{cm}.$$

Example 7.27 Determining the Buoyancy Force Acting on Core in Casting of an Alloy

An aluminum alloy (96 wt% *Al*-4 wt% *Cu*) component is cast in a sand mold using a sand core that weighs 15 kgf. Determine the buoyancy force tending to lift the core during pouring. The densities of aluminum and copper are 2.7 and 8.96 g/cm^3, respectively.

Solution

wt% $Al = 96$, wts% Cu = 4, $\rho_{Al} = 2.7$ g/cm^3, $\rho_{Cu} = 8.96$ g/cm^3, $\rho_s = 1600$ kg/m^3, $F_b = ?$

The density of the alloy can be determined by (Huda, 2020):

$$\rho_{\text{alloy}} = \frac{100}{\dfrac{\text{wt\%Al}}{\rho_{Al}} + \dfrac{\text{wt\%Cu}}{\rho_{Cu}}} = \frac{100}{\dfrac{96}{2.7} + \dfrac{4}{8.96}} = \frac{100}{35.55 + 0.446}$$

$$= 2.778\,\text{g}/\text{cm}^3 = 2778\,\text{kg}/\text{m}^3$$

$$\rho_s = \frac{\text{Mass of core}}{\text{Volume of core}}$$

$$V = \text{Volume of core} = \frac{\text{Mass of core}}{\rho_s} = \frac{15}{1600} = 0.009375\,\text{m}^3$$

By using Eq. 7.11

$$F_b = g \cdot V \cdot \left(\rho_{\text{alloy}} - \rho_s \right) = 9.81 \times 0.009375 \times \left(2778 - 1600 \right) = 108.34\,\text{N}$$

The buoyancy force acting on the core $= 108.34$ N.

Questions and Problems

7.1. Complete the following statement(s) by picking up the most appropriate answer(s).

   (1) The metallic pieces that are used to support cores in position, are called ______.

      (a) Core prints, (b) chaplets, (c) core box, (d) choke

   (2) The system comprising of sprue, runner, and riser, is called __________.

(a) Casting system, (b) pouring system, (c) mold system, (d) gating system

(3) The Reynolds number, in casting design, value in the range of 0–2000 indicates __________.

(a) Laminar flow, (b) turbulence flow, (c) transition flow, (d) abrupt flow

(4) The shrinkage cavity defect in sand casting is mainly caused by poor __________.

(a) Sprue design, (b) runner design, (c) riser design, (d) choke design

(5) The use of a straight sprue, in sand casting, will most likely cause ____________.

(a) Shrinkage cavity defect, (b) crack defect, (c) air entrainment defect, (d) no defect

(6) The taper angle in the faces of a pattern, for its easy removal, is called __________.

(a) Draft allowance, (b) shrinkage allowance, (c) machining allowance, (d) finishing allow

7.2. Draw a labeled sketch of a sand mold showing the gating system, parting line, core, chaplet, and mold cavity.

7.3. Justify the taper in sprue design by the application of *law of continuity*.

7.4. State the rules of good sprue design.

7.5. Why is a cylindrical riser design generally preferred as compared to rectangular riser?

7.6. Why are chaplets used above and beneath the core in a sand mold?

P-7.7. The 3D sketch with dimensions of an aluminum component is shown in Fig. 7.20. This component is to be sand cast. Determine the wood pattern dimensions for the casting.

Fig. 7.20 The sketch of aluminum component for P-7.7

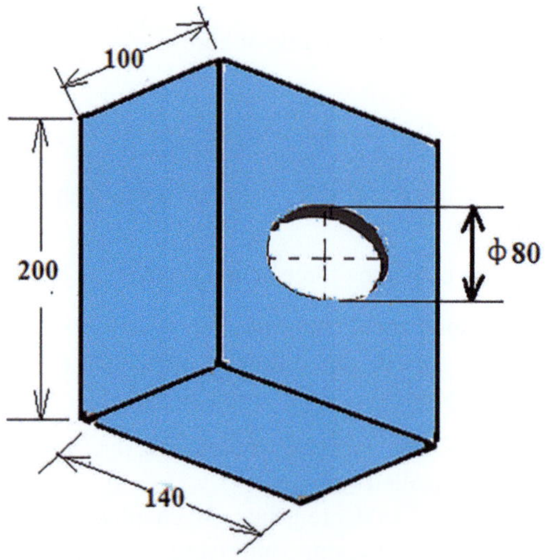

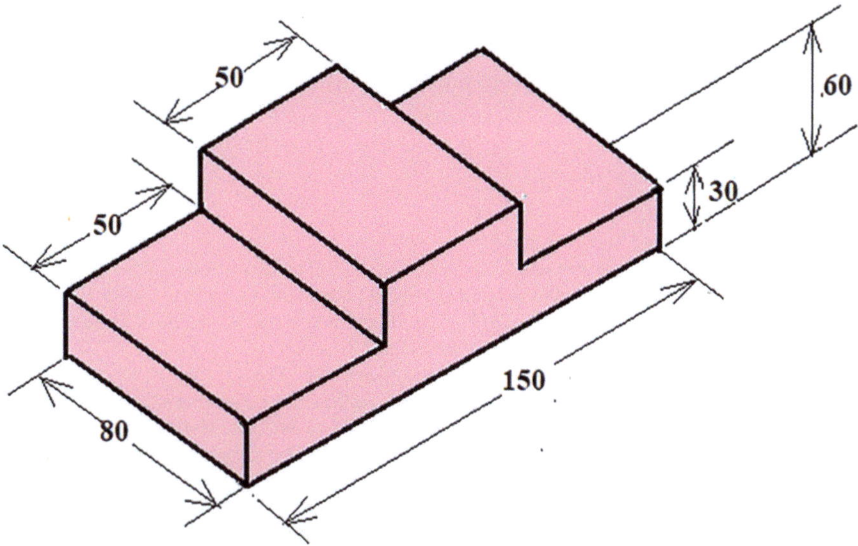

Fig. 7.21 Sketch of the pattern with dimensions for P-7.15

P-7.8. The height of a sprue is 8 cm, and its cross-sectional areas at the top and the bottom are 4.0 and 2.8 cm^2, respectively. Determine the velocity of molten metal at the bottom of the sprue.

P-7.9. By using the data in P-7.8, determine the velocity of molten metal at the top of the sprue.

P-7.10. The molten metal flow rate, in a sand mold, is 3.0×10^{-4} m^3/s. Determine the time to fill the mold cavity, if the volume of the casting is 0.007 m^3.

P-7.11. Molten metal was poured into a 20-cm-long sprue at a volumetric flow rate of 40 cm^3/s. Design the sprue by determining its base diameter in order to keep the same flow rate.

P-7.12. A 9-cm-long sprue is designed for casting a metal component. The density (liq.) of the metal is 7.05 g/cm^3. The viscosity of the liquid metal is 0.01 Pa s. The channel diameter is 0.4 cm. Predict the type of flow (laminar, transition, turbulent). Will this sprue design cause air-entrainment/dross defect?

P-7.13. Design a side riser for a sand mold for the casting with dimensions: 25 cm × 20 cm × 12 cm. Take the height of the riser equal to its diameter.

P-7.14. The volume of a sand core, located inside a mold cavity, is 1500 cm^3. The core is used in the sand casting of an aluminum component. Determine the net buoyant force on the core. The density of aluminum (liq) is 2.4 g/cm^3, and the sand density is 1.6 g/cm^3.

P-7.15. The pattern dimensions for a steel component are shown in Fig. 7.21. The component is required to be made by sand casting. Assume that the parting line is located 30 mm above the base of the pattern. Each flask dimensions are

330 mm × 330 mm × 200 mm. The density of steel (liq) is 7.86 g/cm^3, and the density of sand is 1.65 g/cm^3. Determine the weight (in kgf) needed to be placed on the top of the cope to prevent its lifting.

References

Charbauski, D. (2024). *Understanding the gating system.* American Foundry Society. Online: https://www.castingsource.com/column/2024/08/06/understanding-gating-system

Dojka, R., Jezierski, J., & Szucki, M. (2022). The importance of the geometry of the down-sprue in the gravity casting process. *Materials (Basel), 15*(14), 4937.

Huda, Z. (2020). *Metallurgy for Physicists and Engineers.* CRC Press.

Jiang, R., Wu, G.-h., Xie, H., et al. (2024). Linear contraction of sand casting Mg–9Gd–3Y–0.5Zr alloy. *Transactions of Nonferrous Metals Society of China, 34*(5), 1441–1455.

Sama, S. R., Badamo, T., Lynch, P., & Manogharan, G. (2019). Novel sprue designs in metal casting via 3D sand-printing. *Additive Manufacturing, 25*, 563–578.

Shuvo, M. M., & Manogharan, G. (2021). Novel riser designs via 3D sand printing to improve casting performance. *Procedia Manufacturing, 53*, 500–506.

Zarubin, A. M., & Zarubina, O. A. (2017). Controlling the flow rate of melt in the mold when chill mold casting of aluminum alloys. *Special Methods of Casting Litiejnoje Proizwodstwo, 6*, 16–20.

Chapter 8
Worked Casting Design Project

8.1 Project Description

Assume that you are an engineering manager in a foundry. You have been requested by your client to manufacture, by sand casting, the aluminum component, as shown in Fig. 8.1. Design the pattern, the gating system, and the mold for the sand casting. Your casting design should include the following aspects:

- Design a wood pattern by calculating the changes in the nominal dimensions of your component and by adding the machining allowance and the shrinkage allowance. Draw a sketch of the pattern.
- Design the sprue by specifying the sprue dimensions. Make sprue-design calculations to verify laminar flow of molten metal, to determine a suitable volumetric flow rate, and the minimum time to fill the mold cavity.
- Design the riser to ensure that the metal in riser solidifies last as compared to the main casting.
- Design the mold and determine the metallostatic force.

8.2 Project Objectives

The aim of this project is to design a sand casting system for manufacturing the aluminum component (Fig. 8.1). The following are the project objectives:

- To design a wood pattern by calculating the changes in the nominal dimensions of the component and by adding the machining allowance and the shrinkage allowance. It is required to draw a sketch and the drawing of the pattern.
- To design the sprue by specifying its dimensions, and to ensure that the sprue design is appropriate to achieve laminar flow of molten metal.

© The Author(s), under exclusive license to Springer Nature Switzerland AG 2025

Z. Huda, *Metal Casting Engineering*, Mechanical Engineering Series, https://doi.org/10.1007/978-3-031-84620-5_8

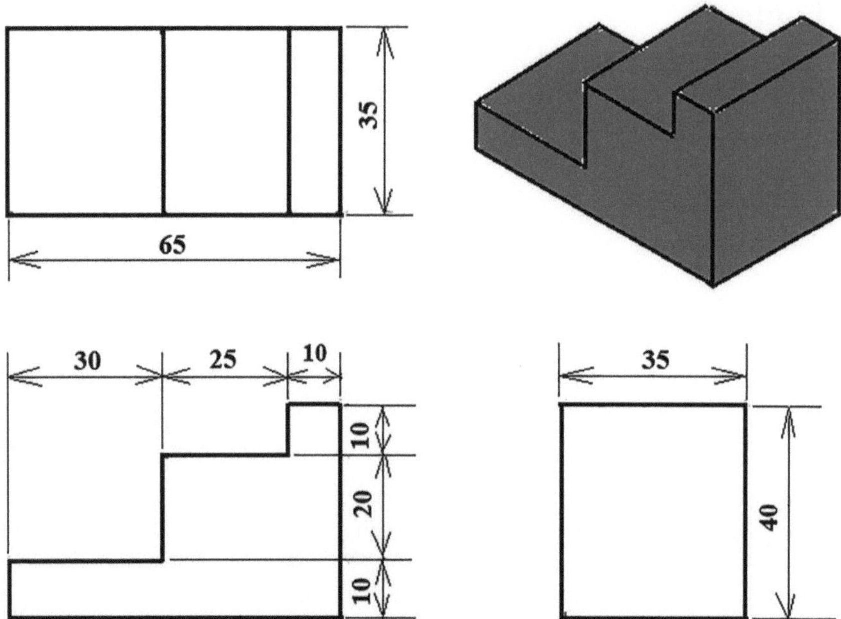

Fig. 8.1 The sketch and drawing of the aluminum component (dimensions in mm)

- To determine the volumetric flow rate and the minimum time to fill the mold cavity.
- To design the riser to solidify last as compared to the main casting.
- To design the mold and to calculate the metallostatic force.

8.3   Solution to the Casting Design Project

8.3.1   Pattern Design

In order to design the pattern for the casting, it is necessary to specify the parting line and the parts that would be in cope and drag in the sand mold. Figure 8.2 illustrates the mold sketch showing the parting line and the relevant dimensions of the casting (see also Fig. 8.1).

First, we refer to the data in Table 7.1. Since all the dimensions in Fig. 8.1 are under 75 mm, the machining allowance for the surface wend the cope dimensions (each), for aluminum (a non-ferrous metal), is 1.6. Additionally, the shrinkage allowance for aluminum may be taken as 1%. Hence, the pattern dimensions can be calculated as illustrated in Table 8.1.

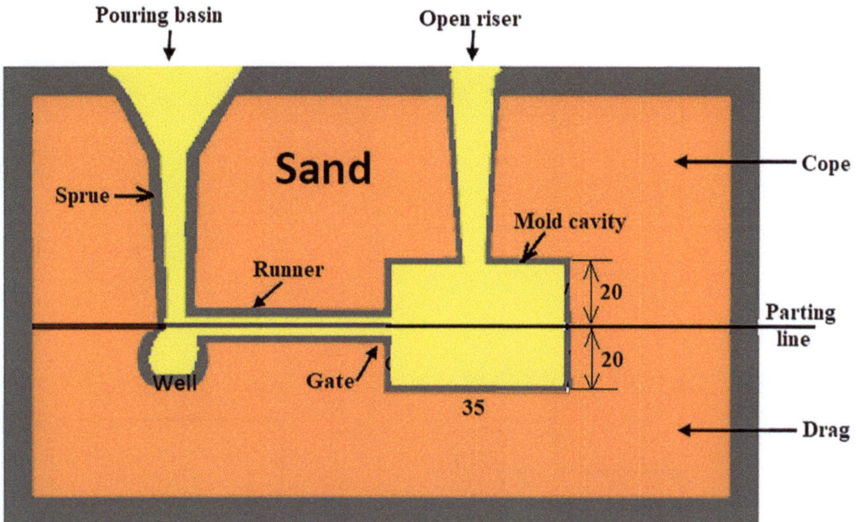

Fig. 8.2 The casting dimensions in the sand mold

Table 8.1 Pattern dimensions (in mm) calculation for the casting design project

Original dimension	Machining allowance	Shrinkage allowance	Pattern dimension
10	1.6 (cope) = 1.6	$(1\%) * (10 + 1.6) = 0.116$	11.716
20	1.6 (cope) + 1.6 (surface) = 3.2	$(1\%) * (20 + 3.2) = 0.232$	23.432
10	1.6 (surface) = 1.6	$(1\%) * (10 + 1.6) = 0.116$	11.716
35	1.6	$(1\%) * (35 + 1.6) = 0.366$	36.966
30	1.6	$(1\%) * (30 + 1.6) = 0.316$	31.916
25	1.6	$(1\%) * (25 + 1.6) = 0.266$	26.866
10	1.6	$(1\%) * (10 + 1.6) = 0.116$	11.716

Based on the data in Table 8.1, the pattern dimensions for the casting are illustrated in Fig. 8.3. The pattern dimensions in the sand mold are shown in Fig. 8.4.

8.3.2   Sprue Design

The sketch of the sand mold with pattern dimensions (Fig. 8.4) indicates that the height of the cope (or the sprue length) should be in the range of 60–100 mm (h = 60–100 mm). Since a high h value yields a high base velocity of molten metal, it is appropriate to take h = 80 mm = 0.08 m. Accordingly, the base velocity of molten metal can be calculated, as follows:

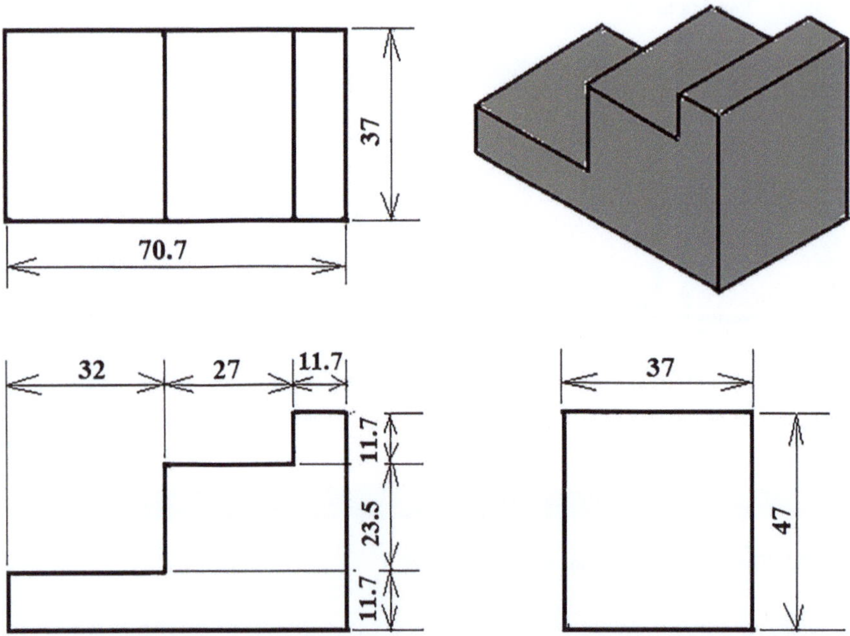

Fig. 8.3 The sketch and drawing showing the pattern dimensions in mm (see the data in Table 8.1)

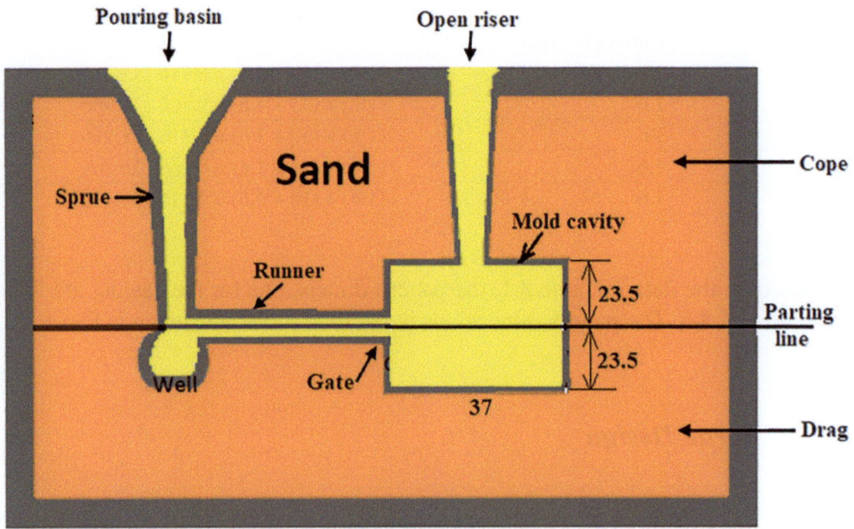

Fig. 8.4 The pattern dimensions in the sand mold

$$v = \sqrt{2 \cdot g \cdot h} = \sqrt{2 \times 9.81 \times 0.08} = 1.25 \, \text{m/s}$$

The physical properties data for liquid aluminum are as follows:

Density of liquid aluminum $= \rho = 2357 \, \text{kg/m}^3$ Viscosity $= \mu = 1.3 \times 10^{-3} \, \text{Pa} \cdot \text{s}$

Now, the objective of the sprue design is to ensure laminar flow, which necessitates $R_e < 2000$.

Taking $R_e = 1900$, we can use Eq. 7.6, as follows:

$$\frac{v \cdot D \cdot \rho}{\mu} = R_e = 1900$$

$$\frac{1.25 \times D \times 2357}{1.3 \times 10^{-3}} = 1900$$

$$2946.25 \, D = 2.47$$

$$D = 0.00083828 \, \text{m} = 0.838 \, \text{mm}$$

It means that laminar flow of the molten metal will be ensured by designing the gating system or sprue, as follows:

Channel diameter = sprue base diameter $= D = 0.84 \, \text{mm}$, sprue length $= h = 80 \, \text{mm}$.

However, there are practical difficulties in creating so small channel diameter (D <1 mm). It is, therefore, better to make a second attempt by fixing $D = 1$ mm $= 0.001$ m and taking $h = 0.06$ m.

$$v = \sqrt{2 \cdot g \cdot h} = \sqrt{2 \times 9.81 \times 0.06} = 1.085 \, \text{m/s}$$

$$R_e = \frac{v \cdot D \cdot \rho}{\mu} = \frac{1.085 \times 0.0010 \times 2357}{1.3 \times 10^{-3}} = 1967$$

Since $R_e < 2000$, the flow of molten metal is **laminar**. The final sprue design is as follows:

Sprue base diameter $= D = 1 \, \text{mm}$, sprue length $= h = 60 \, \text{mm}$.

8.3.3   Determination of the Volumetric Flow Rate and the Time to Fill the Mold Cavity

In order to determine the volumetric flow rate, it is necessary to first find the sprue base area.

$$\text{Sprue base area} = A_2 = \frac{\pi}{4}D^2 = \frac{\pi}{4}\times 1.0^2 = 0.785\,\text{mm}^2$$

$$\text{Volumetric flow rate} = Q = v_2 \cdot A_2 = 1085\times 0.785 = 852.16\,\text{mm}^3\,/\,\text{s}.$$

The volume of the mold cavity can be determined by reference to the pattern drawing (Fig. 8.3).

$$\text{Volume of the mold cavity} = V = V_1 + V_2 + V_3$$

$$V = \left(46.86\times 37\times 11.72\right) + \left(32\times 37\times 26.866\right) + \left(11.72\times 37\times 32\right)$$

$$V = 20,320 + 31,809.34 + 13,876.5 = 66,005.84\,\text{mm}^3$$

By using Eq. 7.5,

$$T_{\text{mf}} = \frac{V}{Q} = \frac{66,005.84}{852.16} = 77\,\text{s} = 1.3\,\text{min}$$

The minimum time to fill the mold cavity is 1.3 min, which is reasonably short and acceptable.

8.3.4   Riser Design

The design of riser requires the determinations of the modulus of casting (M_c) and the modulus of riser (M_r). The volume of casting V_c and the surface area A_c were calculated by *SolidWorks*.

$$V_c = 50,750\,\text{mm}^3, A_c = 10,250\,\text{mm}^2$$

$$M_c = \frac{V_c}{A_c} = \frac{50,750}{10,250} = 4.95$$

By using Eq. 7.8,

$$M_r = 1.2M_c = 1.2\times 4.95 = 5.94$$

It is intended to use a top riser in the sand mold. For a top riser, $H = 2D$.
By using Eq. 7.9,

$$M_{\text{cylinder}} = \frac{D\cdot H}{2\times\left(D+2H\right)} = M_r = 5.94$$

$$\frac{D\cdot\left(2D\right)}{2\times\left(D+2\times 2D\right)} = 5.94$$

$$\frac{2D^2}{10D} = \frac{D^2}{5D} = 5.94$$

$$D^2 = (5D) \times 5.94 = 29.7D$$

$$D = 29.7\,\text{mm}$$

$$H = 2D = 2 \times 29.7 = 59.4\,\text{mm}$$

The riser design specifications: diameter = 30 mm, height = 60 mm.

8.3.5   *Mold Design:* **Metallostatic Force**

The design of the sand mold is illustrated in Fig. 8.5.

h = sprue height = 60 mm, c = 23.5 mm, $H = h - c = 60$–23.5 = 36.5 mm = .0365 m, g = 9.81 m/s^2, A_p = 37 mm × 70.7 mm = 0.037 m × 0.071 m = 2.627 m^2, ρ_m = 2357 kg/m^3

By using Eq. 7.14,

$$F_m = A_p \cdot g \cdot \rho_m \cdot H = 0.002627 \times 9.81 \times 2357 \times 0.0365 = 2.217\,\text{N}$$

The metallostatic force = F_m = 2.217 N.

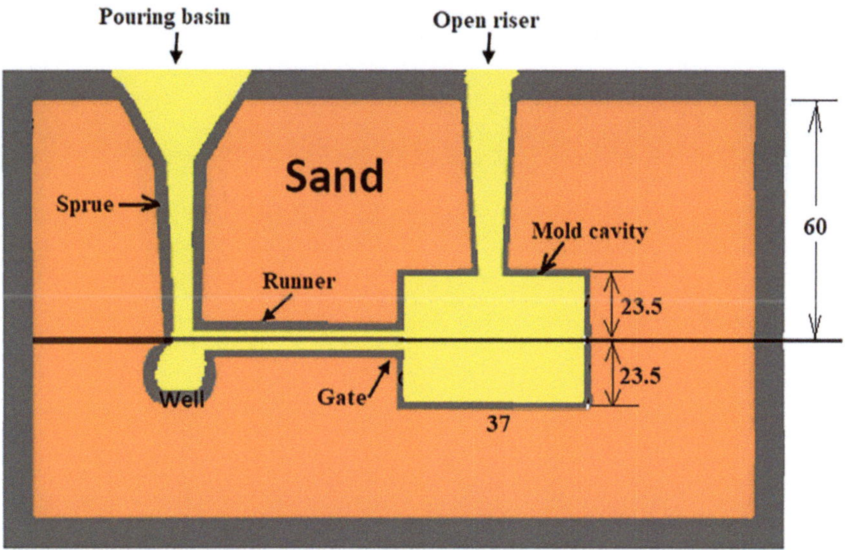

Fig. 8.5 The mold design for casting the aluminum component showing the sprue height an the pattern dimensions in the cope and the drag

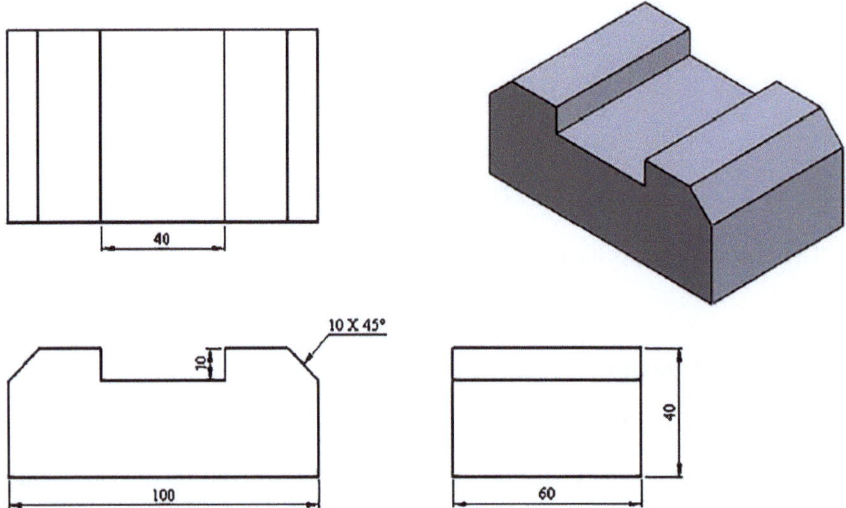

Fig. 8.6 Component drawing for Project # 8.2

Exercise Projects

8.1. Based on the mold design of the project (Sect. 8.3.5), determine the weight required to prevent the lifting of the cope due to the metallostatic force.

8.2. Design the pattern and the gating system for sand casting of the component shown in Fig. 8.6. Choose an appropriate metal to ensure laminar flow.

Chapter 9
Expendable-Pattern Expendable-Mold Casting Processes

Nomenclature

α	Coefficient of linear (thermal) expansion
l_o	Original length
Δl	Change in length
T_o	Ambient temperature
T	Processing temperature
σ	Flexural strength of the shell bar
P	Load at which failure occurs
H	Height of the cross-section of the shell bar
$\sigma_{h(\max)}$	Maximum hoop stress acting on the mold wall (shell)
p_i	Internal pressure that caused bursting
r_o	Outer radius of the cylindrical mold
r_i	Inner radius of the cylindrical mold
M	Modulus
U_{MF}	Av. mold filling velocity of molten metal
v_d	Vacuum degree
ρ	Density
h	Sprue height
t_p	Pouring time
K_t	Correction factor
m	Total mass of the molten metal passing through the ingate
A_c	Minimum choke area
μ	Flow-lost factor
v	Maximum metal front velocity
h_z	Heat transfer coefficient at metal/foam interface (J/s·cm^2·K)
$C_{p,z}$	Specific heat of the foam-pattern material (polystyrene) (J/g·K)
ρ_z	Density of the foam
T_f	Initial foam-pattern temperature (°C)
T_m	Av. metal front temperature (°C)
T_z	Kinetic zone temperature

Z. Huda, *Metal Casting Engineering*, Mechanical Engineering Series, https://doi.org/10.1007/978-3-031-84620-5_9

Abbreviations

CSA	Cooling surface area
EPEM casting	Expendable-pattern expendable-mold casting
EPS	Expanded polystyrene
FMC	Full mold casting
ICP	Investment casting process
LFC	Lost foam casting

9.1 What Are Expendable-Pattern Expendable-Mold Casting Processes?

Expendable-pattern expendable-mold casting (EPEM) casting processes involve the use of temporary patterns with single-use mold. The examples of EPEM casting processes include investment casting, lost foam casting, and full mold casting. In EPEM casting processes, an expendable pattern is first made from a low-melting point material (e.g. wax, foam, etc.) and the mold is built around it. The expendable pattern is then melted or burnt out as the molten metal is poured in. Finally, the mold is broken away to retrieve the casting.

9.2 Investment Casting Process

9.2.1 What Is Investment Casting?

The *investment casting process* (ICP) involves the preparation of a wax pattern, which is then coated with thick slurry containing ceramic particles. Once the slurry has dried, it is fired in an oven. The firing process hardens the ceramic shell and melts out the wax, leaving behind a hollow ceramic mold. The metal is then poured into the mold, which is destroyed after the metal has solidified and cooled.

9.2.2 Advantages and Limitations of ICP

As a method of producing high-quality casting, the investment casting process (*ICP*) provides the highest degree of dimensional accuracy of all casting processes and plays a unique role in producing products with complex shapes (Liu et al., 2015; Yang & Park, 2009). Investment casting can produce parts with high accuracy with tight tolerances. The *ICP* can also produce precise parts with excellent surface finish; thus, the need for additional surface finishing is greatly reduced. The ICP enables us to cast highly complex parts which do not require any draft, unlike other casting processes. It is possible to cast a wide variety of metals and alloys of varying melting points (e.g. steel, superalloy, aluminum, bronze, cobalt, brass, etc.). The *ICP* is used to manufacture gas-turbine blades, assembled disk parts, nuclear reactor parts, and the like (Pattnaik et al., 2012).

There are a few limitations of investment casting. The ICP involves many steps; each casting cycle takes much longer to complete. It means that investment casting is a time-consuming process. Investment casting is quite costly compared to other casting processes, especially with a low volume of production. It is difficult to cast parts that require cores. In particular, very small holes would be challenging or impossible to cast using the investment casting process.

9.2.3   The Steps in Investment Casting Process

The investment casting process mainly includes preparing wax patterns, coating multilayer ceramic shells, de-waxing, sintering shells, pouring, removing ceramic shells, and post-treatment. The following principal steps are involved in investment casting process: (1) die making, (2) wax injection, (3) slurry coating and stuccoing, (4) de-waxing by heating, (5) firing, and (6) pouring of molten metal (see Fig. 9.1).

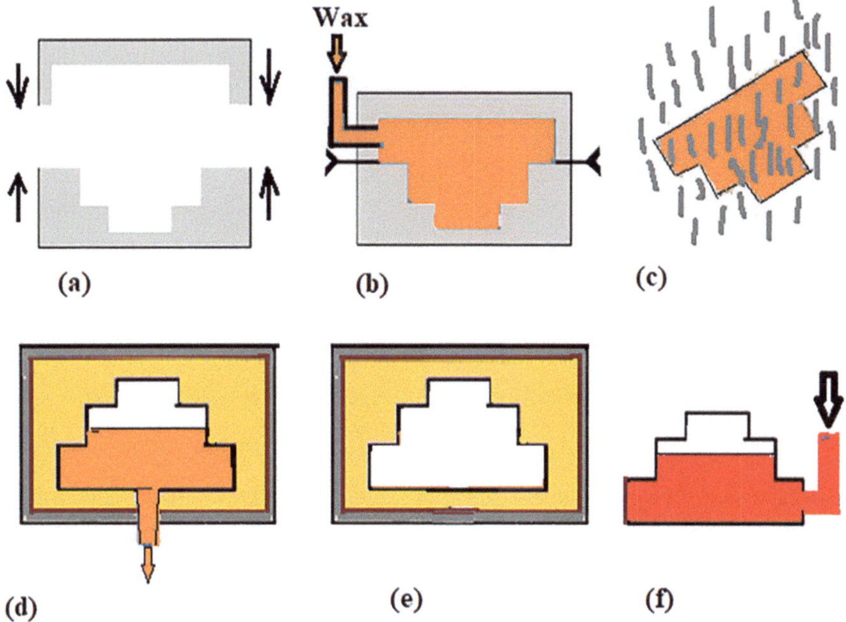

Fig. 9.1 Principal steps in investment casting process. (**a**) Die making. (**b**) Wax injection. (**c**) Slurry coating and stuccoing. (**d**) De-waxing by heating. (**e**) Firing. (**f**) Pouring of molten metal

In addition to these steps, there are other steps, such as the removal of the ceramic shell, inspection, and quality control. These operational steps are briefly explained in the following paragraphs.

The first step in investment casting is to design and construct a metal die (Fig. 9.1a). This die creates a wax replica of the desired part. The recent trend is to replace costly die making by 3D printing of wax (see Sect. 9.2.4). Second, molten wax is injected under high pressure into the die cavity (Fig. 9.1b). Once the wax pattern has been created, it is usually assembled with other components to create a gate and runner metal delivery to create the casting. A feeder system is provided to the wax pattern, which has many of the same functions as a sprue and riser system that is used for sand casting processes. The feeder, in ICP, holds extra molten metal during the pour process. After the molten metal has filled the mold and begun to cool, shrinkage and contraction may occur to leave voids. The extra molten metal in the feeder will fill the voids in the ceramic mold after this shrinkage to ensure the cavity is entirely filled. The next step is to dip the wax pattern assembly in a ceramic slurry, and then, the assembly is covered with sand stucco and given time to dry (see Fig. 9.1c). If necessary, cycles of wet dipping and stuccoing are repeated until the desired shell thickness is achieved. As the ceramic shell gets dried, it becomes strong enough to retain molten metal during the casting process.

Slurry coating and stuccoing is followed by de-waxing (Fig. 9.1d). De-waxing involves putting the assembly inside a steam autoclave to melt away the wax. Heating or firing is continued until all the wax is burned out. Once all the wax has been removed, the cavity in the shape of a desired cast part is obtained (Fig. 9.1e). Then, molten metal is poured into the mold cavity to produce the casting (Fig. 9.1f). When the casting has cooled, the mold shell is broken off from the casting. If necessary, the gates and runners are cut from the casting. Although the surface finish is generally good, in some cases, it is necessary to do sandblasting, grinding, and machining to finish the casting.

9.2.4   Recent Advances in ICP

9.2.4.1   3D Printing of Wax Pattern

In traditional ICP, the sacrificial wax patterns are created by injection molding wax, and as such require the use of metal tooling; the substantial cost and time required to generate wax pattern tooling limits the range of applications for which investment casting is economically competitive. Modern ICP is based on the design freedom and cost savings of direct 3D printing master patterns. Thus, stereo-lithography 3D printing of wax offers a low-cost, high-yield alternative to traditional ICP.

Recently (2024), Mukhtarkhanov and co-researchers have proposed a novel method for the evaluation of expansion forces that occur during the heating of investment-casting waxes with the help of a rheometer. They tested three types of commercially available waxes and one 3D-printable wax. It was found that the maximum normal load produced by the Wax-3D sample was equal to 36.3 N, which is considerably higher compared to loads generated by the IC wax (18.8 N). It was discovered that the 3D-printed version of the Wax3D material is preferable to its cast counterpart when considering the investment casting applications. This is due to the higher porosity of the 3D-printed wax, which partially gives room for expansion during de-waxing (Mukhtarkhanov et al., 2024).

9.2.4.2 Shell Thickness Uniformity

Recently (2024), Dong and co-researchers have investigated the effects of gas-turbine blade casting system design on ceramic-shell thickness uniformity and its correlation with casting dimensional accuracy. Their findings reveal that strategic adjustments to the wax pattern attitude angle and module diameter can significantly enhance shell thickness uniformity, thereby potentially increasing the structural integrity and dimensional precision of turbine blades (Dong et al., 2024)

9.2.4.3 Investment Casting of High-Performance Aluminum Alloys

The investment casting process provides great potential to achieve excellent surface finishes and complex geometry for aluminum alloy components; however, these high-performance aluminum alloys are almost impossible to be investment cast due to their hot cracking susceptibility and severe shrinkage during solidification. Recently (2024), Chi and Li have introduced nanotechnology to improve the processability of high-performance wrought aluminum alloys in investment casting by adding a small volume fraction of nanoparticles into the aluminum alloys. This work showed the unprecedented success of nanotechnology-enabled investment casting of high-strength wrought aluminum alloys (AA6061, AA2024, and AA7075) for excellent mechanical properties (Chi & Li, 2024).

9.2.5 Engineering Analysis of ICP

9.2.5.1 Analysis for Thermal Expansion of Wax

The thermal expansion of wax in ICP is technologically important. The coefficient of linear expansion for wax can be experimentally determined by:

$$\alpha = \frac{\Delta l}{l_o * (T - T_o)}$$

(9.1)

where α is the coefficient of linear expansion (/°C), l_o is the original length, Δl is the change in length, T_o is the ambient temperature (°C), and T is the processing temperature (°C) (see Example 9.1).

9.2.5.2 Analysis for Flexural Strength and Max. Hoop Stress of Ceramic Shell

The mechanical properties of a ceramic shell can be experimentally determined by preparing bars with a size of $L \times w \times$ thickness of the shell. To prepare the shell, a pre-mixed colloidal silica ceramic slurry can be used as a binder. As for stucco material, a fused silica may be applied upon each layer of slurry except for the last sealing coat. Each layer of the coat is dried for a specified time duration (Mukhtarkhanov et al., 2024). The flexural strength of the ceramic shell can be determined by:

$$\sigma = \frac{1.5 * P * L}{w * H^2} \tag{9.2}$$

where σ is the flexural strength of the shell bar (MPa), P is the load at which failure occurs (N), L is the length (mm), w is the width (mm) and H is the height of the cross-section of the shell bar (mm) (see Example 9.2).

The burst test has been prescribed as an effective method for measuring the integrity of ceramic shells (Babu et al., 1990). During the burst test, the compressed air is introduced to the cylindrical ceramic mold, and the bursting pressure is recorded. By knowing the internal (bursting) pressure and geometrical characteristics of the ceramic mold, the hoop stress can be calculated by (Wibawa et al., 2020):

$$\sigma_{h(\max)} = p_i * \left(\frac{r_o^2 + r_i^2}{r_o^2 - r_i^2} \right) \tag{9.3}$$

where $\sigma_{h(\max)}$ is the maximum hoop stress acting on the mold wall (shell) (MPa), p_i is the internal pressure that caused bursting (MPa), and r_o and r_i are the outer and inner radii of the cylindrical mold, respectively (see Example 9.3).

9.2.5.3 Analysis for Feeder Positioning in ICP

It is a well-known metallurgical phenomenon that when an alloy is poured into a mold, it starts to shrink or contract in volume as it cools down to liquidus temperature and subsequently solidifies. The two stages of volumetric contraction are compensated for by adding feeders to the mold cavity design. But unlike sand casting, in investment casting, the gating/feeding elements provide dual or combined function of both (Chalekar et al., 2015). In order for a casting to be free from shrinkage defect, the feeder, in ICP, is positioned at the slowest freezing section, i.e. the section having the maximum modulus. The modulus M can be determined by:

$$M = \frac{V}{CSA} \tag{9.4}$$

where V is the volume of the casting, and CSA is the cooling surface area of the casting (the surface area excluding the surfaces not facing the mold) (see Examples 9.4 and 9.5).

9.3   Lost Foam Casting

9.3.1   Working Principle and Advantages of Lost Foam Casting

Lost foam casting (LFC) is a relatively recent casting technology that exhibits characteristics of both investment casting and sand casting. LFC is a form of evaporative pattern casting where the pattern is made of foam. The foam pattern is coated with a refractory compound to form a shell; the "shell" is then surrounded by compacted un-bonded sand to give it strength. When molten metal is poured into the shell, the foam pattern evaporates on contact. The LFC process delivers the quality and accuracy of investment casting with lower costs and greater flexibility in sand casting.

One of the problems that is encountered in the traditional sand casting process is the removal of pattern after sand is compacted around it. Since the mold is destroyed when the pattern is removed, the resulting mold cavity becomes inaccurate, which in turn results in a product with inaccurate dimensions and undesired shape. This problem is overcome by LFC by replacing the solid pattern by polystyrene (foam), which evaporates when approached by molten metal. The vapors of the polystyrene leave the cavity through the pores of the mold and there is no pattern-removal step. This results in producing a part with better accuracy and a smoother surface finish (Tolcha & Tibba, 2014).

9.3.2   Steps in Lost Foam Casting

The following are the steps in the lost foam casting (LFC) process.

Step I. A foam (polystyrene) pattern is prepared (Fig. 9.2a).
Step II. The foam pattern is coated with a refractory compound by use of a spray gun (Fig. 9.2b).
Step III. The foam pattern is placed in a mold box, and sand is compacted around the pattern by use of a compactor (Fig. 9.2c).

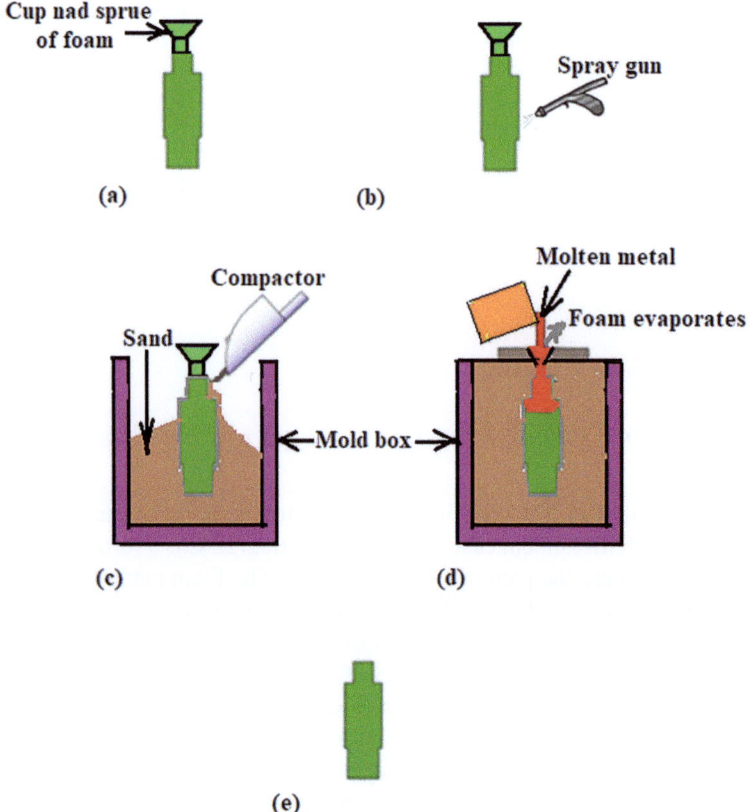

Fig. 9.2 The steps in lost foam casting process. (**a**) Foam pattern. (**b**) Spraying refractory compound. (**c**) Sand is compacted around the pattern. (**d**) Pouring of molten metal. (**e**) Casting (product)

Step IV. Molten metal is poured into the portion of the pattern that forms the cup and sprue. As the metal is admitted into the mold, the polystyrene foam is vaporized ahead of the advancing liquid, thereby allowing the resulting mold cavity to be filled (Fig. 9.2d).

Step V. The cup and sprue are cut off to obtain the casting (Fig. 9.2e).

9.3.3 Engineering Analysis of Lost Foam Casting

9.3.3.1 Determination of Mold Filling Velocity in LFC

The mold filling velocity, in the LFC, is quite different from the conventional empty mold (gravity) casting. The velocity at the melt front is restricted by foam degradation, gas elimination, pressure build-up in the mold, and the like. It is affected by the original process parameters, such as the degree of vacuum, the pattern density, and the metallic static head. The influence of the gravity on the mold filling velocity, in LFC, is minor as compared to conventional empty mold casting (Li et al., 2020). The average mold filling velocity of cast iron, in LFC, is related to the down–sprue height h and other variables by the following formula (Li et al., 2013):

$$U_{\mathrm{MF}} = 661.967 + 153.96 * v_{\mathrm{d}} - 0.7632 * \rho + 0.8074 * h$$
$$- 0:940321 * T + 0:0003397 * T^2 \tag{9.5}$$

where U_{MF} is the average mold filling velocity of molten cast iron (cm/s), v_{d} is the degree of vacuum (–MPa), ρ is the pattern density (g/L), h is the sprue height (cm), and T is the pouting temperature (°C) (see Example 9.6).

9.3.3.2   Determination of Pouring Time and Choke Area in LFC

The pouring time for small and medium-sized parts, in lost foam casting, can be determined by (ZHYcasting, 2022):

$$t_{\mathrm{p}} = K_t \cdot \left(\sqrt[3]{m} + \sqrt{m} \right) \tag{9.6}$$

where t_{p} is the pouring time (s), K_t is the correction factor ($K_t = 0.85$), and m is the total mass of the molten metal passing through the ingate (kg) (see Example 9.7). Once the pouring time has been calculated, the minimum choke area can be determined by:

$$A_{\mathrm{c}} = \frac{m}{0.31 * \mu * t_{\mathrm{p}} * \sqrt{h}} \tag{9.7}$$

where A_{c} is the minimum choke area (cm^2), μ is the flow-lost factor ($\mu = 0.4$–0.6 for iron castings), and average static head height (cm) (see Example 9.8).

9.3.3.3   Determination of the Molten Metal Front Velocity in LFC

The maximum front velocity of molten metal, in LFC, can be determined by (DoE-OSTI, 1999):

$$v = \frac{h_z * \left(T_{\mathrm{m}} - T_z \right)}{C_{p,z} * \rho_z * \left(T_z - T_{\mathrm{f}} \right)} \tag{9.8}$$

where v is the maximum metal front velocity (cm/s), h_z is the heat transfer coefficient at metal/foam interface (J/s•cm^2•K), $C_{p,z}$ is the specific heat of the foam-pattern material (polystyrene) (J/g•K), ρ_z is the density of the foam (g/cm^3), T_{f} is the initial foam-pattern temperature (°C), T_{m} is the av. metal front temperature (°C), and

T_z is the kinetic zone temperature (°C). The kinetic zone temperature can be calculated by:

$$T_z = \frac{T_s + T_e}{2} \tag{9.9}$$

where T_s is the temperature at the foam surface in the kinetic zone (°C), and T_e is the temperature at the edge of the kinetic zone (°C). The significance of Eqs. 9.8 and 9.9 is illustrated in Examples 9.9 and 9.10.

9.4   Full Mold Casting

9.4.1   What Is Full Mold Casting?

The full mold casting (FMC) is an EPEM casting process that combines features of both sand casting and lost foam casting. In FMC, a pattern is made from fused polystyrene beads that are heated in an aluminum die. Once the pattern has cooled, it is coated with a refractory compound, before being placed in a flask and surrounded by sand. A vacuum is applied to the flask and molten metal is poured into the pattern. Molten metal vaporizes the polystyrene material as it flows into the mold cavity. Once cooled, the casting is removed (The Open University, 2017).

9.4.2   Steps in Full Mold Casting Process

The following steps are involved in the FMC process.

1. Polystyrene beads are injected into a heated aluminum die.
2. Steam is injected into the foam pattern, from a steam chest, to fuse and expand polystyrene (EPS) beads.
3. Once the die has cooled, it is opened and the pattern is ejected and stored for 30 days.
4. The pattern is coated with a special coating and dried.
5. The pattern is placed in a flask containing loose, dry sand.
6. More sand is added to the flask and consolidated by vibrations.
7. Vacuum is applied to the flask.
8. The metal is poured.
9. On cooling, the casting is removed.
10. The sand is treated before re-use.

9.4.3   Advantages of FMC

The FMC can save 10–20% due to the simple equipment and low operating cost, as compared to conventional sand casting. It involves low-cost consumables, as the sand binder is eliminated and the sand is recycled. There is no knock-out or de-coring problem, as the loose sand flows away easily. The tool life, in FMC, is excellent, since tooling does not contact either sand or liquid metal. Yields, in FMC, are slightly higher (50–80%) than the conventional methods, since risers are seldom needed. The FMC is capable of producing automotive components, such as camshafts, crankshafts, ball-valve casings, and the like.

9.5   Worked Numerical Examples in EPEMC Processes

Example 9.1 Determining the Thermal Linear Coefficient of Investment-Casting Wax

The $\Delta l/l_o$ ratio for an investment-casting wax at a processing temperature of 35 °C was found to be 2×10^{-3}. Determine the thermal linear coefficient of the wax, if the starting ambient temperature was 10 °C.

Solution

$T_0 = 10$ °C, $T = 35$ °C, $T - T_0 = 35-10 = 25$ °C, $\Delta l/l_o = 2 \times 10^{-3}$, $\alpha =$?

By using Eq. 9.1,

$$\alpha = \frac{\Delta l}{l_o * (T - T_o)} = \frac{2 \times 10^{-3}}{25} 8 \times 10^{-5} \, / °C$$

The linear (thermal) expansion coefficient of the wax $= 8 \times 10^{-5}$/°C.

Example 9.2 Determining the Flexural Strength of Ceramic Shell in ICP

For mechanical testing of the ceramic shell, the bars were prepared with a size of $150 \times 90 \times 5$ mm of the shell. Determine the flexural strength of the ceramic shell, if the average maximum load at break at a layer of slurry is 50 N.

Solution

$P = 50$ N, $L = 150$ mm, $w = 90$ mm, $H = 5$ mm, $\sigma =$?

By using Eq. 9.2,

$$\sigma = \frac{1.5 * P * L}{w * H^2} = \frac{1.5 \times 50 \times 150}{90 \times 5^2} = 5 \, \text{MPa}$$

The flexural strength of the ceramic shell $= 5$ MPa.

Example 9.3 Determining the Maximum Hoop Stress Acting on the Ceramic Shell

A burst test of a ceramic mold was conducted. The dimensions of the cylindrical mold are: inner diameter = 16 mm, length = 85 mm, and shell thickness = 3 mm. The pressure gage showed an internal pressure of 1.48 MPa at bursting. Determine the maximum hoop stress acting on the mold wall (shell).

Solution

Data: d_i = 16 mm, t = 3 mm, p_i = 1.48 MPa, $\sigma_{h(\max)}$ =?

The cross-sectional parameters of the cylindrical mold are shown in Fig. 9.3.

With reference to the cross-section of the cylindrical mold (Fig. 9.3), we can write:

$$d_o = d_i + 2t = 16 + (2 \times 3) = 22 \, \text{mm}$$

$$\text{Outer radius} = r_o = \frac{d_o}{2} = \frac{22}{2} = 11 \, \text{mm}$$

$$\text{Inner radius} = r_i = \frac{d_i}{2} = \frac{16}{2} = 8 \, \text{mm}$$

Data: p_i = 1.48 MPa, r_o = 11 mm, r_i = 8 mm, $\sigma_{h(\max)}$ =?
By using Eq. 9.3,

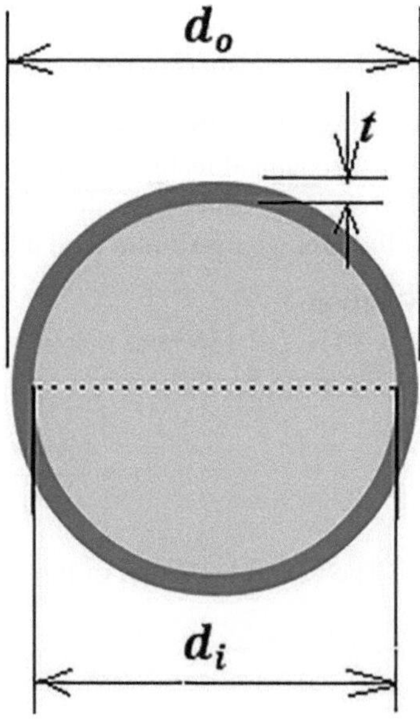

Fig. 9.3 The cross-sectional parameters of the cylindrical mold for Example 9.3

$$\sigma_{h(\max)} = p_i * \left(\frac{r_o^2 + r_i^2}{r_o^2 - r_i^2} \right) = 1.48 \times \left[\frac{\left(11^2 + 8^2\right)}{\left(11^2 - 8^2\right)} \right] = 1.48 \times \left(\frac{121 + 64}{121 - 64} \right)$$

$$= 1.48 \times \left(\frac{185}{57} \right) = 4.8\,\text{MPa}$$

The maximum hoop stress acting on the mold wall (shell) = 4.8 MPa.

Example 9.4 Identifying the Casting Section to Which Feeder for ICP Be Attached

A casting has four sections with the volumes and the cross-sectional area (CSA), as shown in Table 9.1. Determine the section to which the feeder for investment casting should be positioned.

Solution

By using Eq. 9.4 for the data in Table 9.1, the modulus of section 1 is:

$$M = \frac{V}{\text{CSA}} = \frac{70,100}{13,050} = 5.37$$

Similarly, the moduli of other sections can be calculated, which are presented in Table 9.2.

Since the modulus of section 1 ($M_c = 5.37$) is the maximum, the feeder should be attached to section 1 for the investment casting process.

Example 9.5 Determining the Modulus of Feeder for ICP

By using the data in Example 9.4, determine the modulus of feeder to ensure that the feeder is thermally adequate.

Table 9.1 The volume and CSA data for Example 9.1

Section	Volume (mm^3)	CSA (mm^2)
1	70,100	13,050
2	42,180	8250
3	10,600	2700
4	7372	2260

Table 9.2 The volume, CSA, and modulus data for Example 9.1

Section	Volume (mm^3)	CSA (mm^2)	Modulus, M
1	70,100	13,050	5.37
2	42,180	8250	5.11
3	10,600	2700	3.92
4	7372	2260	3.26

Solution

In order for the feeder to be thermally adequate, its modulus should be 1.2 times the modulus of casting, i.e. $M_f = 1.2 * M_c$ (see Eq. 7.8).

Hence, $M_f = 1.2 * M_c = 1.2 \times 5.37 = 6.44$.

Example 9.6 Determining the Mold Filling Velocity of Molten Metal in LFC

The schematic of a lost foam casting process for manufacturing a cast-iron component is shown in Fig. 9.4. The casting process data is given below.

Pattern density = 0.013 g/cm^3, pouring temperature = 1420 °C, vacuum degree = −0.047. Determine the average mold filling velocity of the molten metal.

Solution

$\rho = 0.013$ g/cm^3 = 13.0 g/L, $T = 1420$ °C, $v_d = -0.047$ MPa, $h = 50$ cm (see Fig. 9.4).

By using Eq. 9.5,

$$U_{MF} = 661.967 + 153.96 * v_d - 0.7632 * \rho + 0.8074 * h$$
$$- 0.940321 * T + 0.0003397 * T^2$$

$$U_{MF} = 661.967 + (153.96 \times -0.047) - (0.7632 \times 13) + (0.8074 \times 50)$$
$$- (0.940321 \times 1420) + (0.0003397 \times 1420^2)$$

$$U_{MF} = 661.967 - 7.23612 - 9.9216 + 40.37 - 1335.256 + 684.97$$

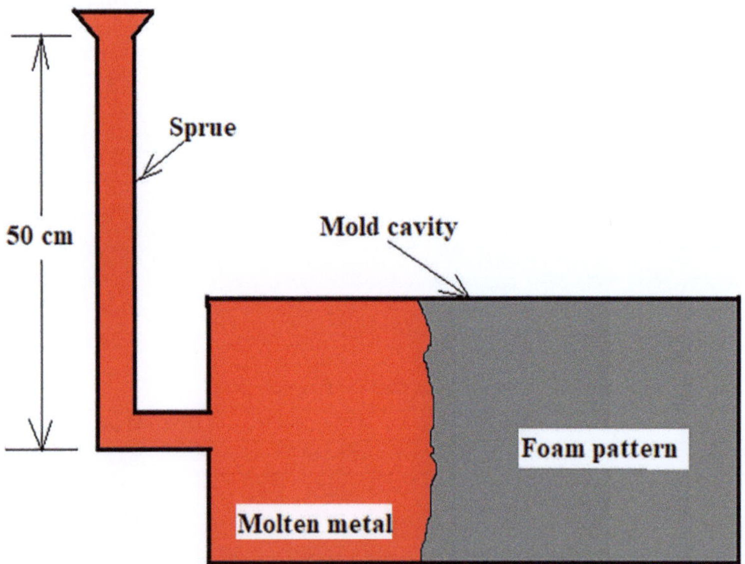

Fig. 9.4 Schematic of LFC process for Example 9.6

$$U_{MF} = 1387.307 - 1352.41 = 34.8 \cong 35 \, \text{cm/s}$$

The average mold filling velocity of the molten metal = 35 cm/s.

Example 9.7 Determination of the Pouring Time in LFC

In an LFC process, the total mass of the molten metal passing through the ingate is 90 kg. Determine the pouring time for the process.

Solution

$K_t = 0.85$, $m = 90$ kg, $t_p = ?$

By using Eq. 9.6,

$$t_p = K_t \cdot \left(\sqrt[3]{m} + \sqrt{m}\right) = 0.85 \times \left(\sqrt[3]{90} + \sqrt{90}\right) = 0.85 \times (4.4814 + 9.4868)$$
$$= 11.87 \, \text{s} \cong 12 \, \text{s}$$

The pouring time in the LFC process = t_p = 12 s.

Example 9.8 Determination of the Choke Area in LFC

By using the data in Example 9.7, determine the minimum choke area for the LFC process, if the average static head height is 45 cm.

Solution

$t_p = 12$ s, $m = 90$ kg, $\mu = 0.5$, $h = 45$ cm, $A_c = ?$

By using Eq. 9.7,

$$A_c = \frac{m}{0.31 * \mu * t_p * \sqrt{h}} = \frac{90}{0.31 \times 0.5 \times 12 \times \sqrt{45}} = \frac{90}{12.477} = 7.213 \, \text{cm}^2$$

The minimum choke area for the LFC process = 7.21 cm^2.

Example 9.9 Determining the Kinetic Zone Temperature

An aluminum alloy component was produced by lost foam casting. The temperature at the foam surface in the kinetic zone is 200 °C, and the temperature at the edge of the kinetic zone is 300 °C. Determine the kinetic zone temperature.

Solution

By using Eq. 9.9,

$$T_z = \frac{T_s + T_e}{2} = \frac{200 + 300}{2} = 250 °C$$

The kinetic zone temperature = T_z = 250 °C.

Example 9.10 Determination of the Maximum Metal Front Velocity

By using the data in Example 9.9, determine the maximum metal front velocity. The following are LFC process data. The heat transfer coefficient at metal/foam interface is 0.07 J/s•cm²•K. The specific heat of the foam-pattern material (polystyrene) is 1.6 J/g•K. The density of the foam is 25.6 kg/m³. The initial foam-pattern temperature is 18 °C. The average metal front temperature is 678 °C.

Solution

$T_z = 250$ °C, $h_z = 0.07$ J/s•cm²•K, $C_{p,z} = 1.6$ J/g•K, $\rho_z = 25.6$ kg/m³ $= 0.0256$ g/cm³, T_f 18 °C, $T_m = 678$ °C, $v = ?$

Solution

By using Eq. 9.8,

$$v = \frac{h_z^{*}\left(T_m - T_z\right)}{C_{p,z}^{*}\,\rho_z^{*}\,\left(T_z - T_f\right)} = \frac{0.07 \times \left(678 - 250\right)}{1.6 \times 0.0256 \times \left(250 - 18\right)} = \frac{29.96}{9.5027} = 3.15\,\mathrm{cm/s}$$

The maximum metal front velocity of the molten metal = 3.15 cm/s.

Questions and Problems

9.1. Encircle the most appropriate answers for each of the following statements.

    (1) Which one is not an evaporative pattern casting process?
     (a) Investment casting, (b) sand casting, (c) lost foam casting
    (2) Which casting process has combined features of LFC and sand casting?
     (a) Full mold casting, (b) investment casting, (c) die casting, (d) V-casting
    (3) Which casting process is unsuitable for complex shapes with excellent surface finish?
     (a) Investment casting, (b) lost foam casting, (c) sand casting, (d) full mold casting.
    (4) Which casting process has been used to manufacture gas-turbine blades?
     (a) Full mold casting, (b) lost foam casting, (c) sand casting, (d) investment casting
    (5) Which casting process involves the use of wax?
     (a) Investment casting, (b) lost foam casting, (c) sand casting, (d) full mold casting.
    (6) Which technology has enabled successful investment casting of high-strength *Al* alloys?
     (a) Nanotechnology, (b) 3D printing technology, (c) aerospace technology
    (7) Which casting process involves the use of polystyrene pattern?
     (a) Full mold casting, (b) investment casting, (c) sand casting, (d) lost foam casting.

9.2. List the principal steps involved in investment casting process (ICP).
9.3. What are the advantages of ICP?
9.4. Is the ICP economical for low volume manufacturing? Explain.
9.5. Why is de-waxing done in ICP? And how is it done?
9.6. What recent development has been made to ensure shell thickness uniformity in ICP?

Table 9.3 The volume and CSA data for P-9.13

Section	Volume (mm³)	CSA (mm²)
1	55,100	14,050
2	42,180	8250
3	62,600	3700
4	7372	2260

9.7. Explain how lost foam casting (LFC) is superior to sand casting in terms of quality.

9.8. Diagrammatically illustrate the steps in LFC.

9.9. Explain full mold casting process.

P-9.10. The length of a wax pattern is 5 cm, and its thermal linear coefficient is 8 $\times 10^{-5}$/°C.

The wax patter is heated from 15 to 40 °C. Determine the final length of the pattern.

P9.11. A burst test of a ceramic mold was conducted. The dimensions of the cylindrical mold are: inner diameter = 14 mm, length = 90 mm, and shell thickness = 2 mm. The pressure gage showed an internal pressure of 1.30 MPa at bursting. Determine the maximum hoop stress acting on the mold wall (shell).

P-9.12. For mechanical testing of the ceramic shell, the bars were prepared with a size of $130 \times 80 \times 4$ mm of the shell. Determine the flexural strength of the ceramic shell, if the average maximum load at break at a layer of slurry is 47 N.

P-9.13. A casting has four sections with the volumes and the CSA, as shown in Table 9.3.

Determine the section to which the feeder for investment casting should be positioned.

Determine, also, the modulus of the feeder to ensure that it is thermally adequate.

P-9.14. In an LFC process, the total mass of the molten metal passing through the ingate is 80 kg. Determine the pouring time for the process.

P-9.15. An aluminum alloy component was produced by lost foam casting. The temperature at the foam surface in the kinetic zone is 240 °C, and the temperature at the edge of the kinetic zone is 320 °C. The heat transfer coefficient at metal/foam interface is 0.07 J/s•cm²•K. The specific heat of the foam-pattern material (polystyrene) is 1.6 J/g•K. The density of the foam is 25.6 kg/m³. The initial foam-pattern temperature is 20 °C. The average metal front temperature is 700 °C. Determine the maximum metal front velocity.

References

Babu, K. M., Jebaraj, P. M., & Chowdiah, M. P. (1990). Standardization of investment casting process using modified binder hydrolysis. *Bulletin of Materials Science, 13*(3), 227–234.

Chalekar, A. A., Somatkar, A. A., & Chinchanikar, S. S. (2015). Designing of feeding system for investment casting process—A case study. *Journal of Mechanical Engineering and Automation, 5*(3B), 15–18.

Chi, Y., & Li, X. (2024). Nanotechnology-enabled rapid investment casting of high-performance wrought aluminum alloys, Manufacturing Letters, 41, 339–343.

DoE-OSTI. (1999). *Advanced lost foam casting technology*. U.S Department of Energy, Office of the Scientific and Technical Information (DoE-OSTI). Retrieved November 16, 2024, from https://digital.library.unt.edu/ark:/67531/metadc685376/m2/1/high_res_d/353173.pdf

Dong, R.-z., Wang, W.-h., Cui, K., et al. (2024). An investigation of ceramic shell thickness uniformity and its impact on precision in turbine blade investment casting. *Journal of Manufacturing Processes, 131*, 507–522.

Li, F. J., Han, J. P., Yi, L. W., et al. (2013). Empirical calculation on the mold filling speed of lost foam iron casting. *AFS Transactions, 121*, 565–568.

Li, F., Zhao, H., Ren, F., et al. (2020). Simulations and experiments of mold filling in lost foam casting. *International Journal of Cast Metals Research, 33*(4–5), 194–200.

Liu, C., Jin, S., Lai, X., & Wang, Y. (2015). Dimensional variation stream modeling of investment casting process based on state space method. *Journal of Engineering Manufacture Proceedings of the IMechE (Part B), 229*(3), 463–474.

Mukhtarkhanov, M., Akayev, S., Gouda, S., et al. (2024). A novel method for evaluating thermal expansion forces during dewaxing of investment casting and 3D-printing waxes. *International Journal of Lightweight Materials and Manufacture* (in press).

Open University. (2017). *Full mold casting (evaporative pattern)*. Retrieved November 16, 2024, from https://www.open.edu/openlearn/science-maths-technology/engineering-technology/manupedia/full-mould-casting-evaporative-pattern

Pattnaik, S., Karunakar, D. B., & Jha, P. K. (2012). Developments in investment casting process—A review. *Journal of Materials Processing Technology, 212*(11), 2332–2348.

Tolcha, M. A., & Tibba, G. S. (2014). Utilization of waste polystyrene material in local foundry technology for manufacturing complex shapes. *International Journal of Engineering Research & Technology, 3*(3), 1975–1979.

Wibawa, L. A. N., Diharjo, K., Raharjo, W. W., & Jihad, B. H. (2020). Stress analysis of thick-walled cylinder for rocket motor case under internal pressure. *Journal of Advanced Research in Fluid Mechanics and Thermal Sciences, 70*(2), 106–115.

Yang, J., & Park, Y. (2009). Scheduling of casting in real foundries using linear programming. *Proceedings of the IMechE, Part B: Journal of Engineering Manufacture, 223*, 1351–1360.

ZHYcasting. (2022). *Determination of pouring size of gating system in lost foam casting process of box*. Retrieved November 15, 2024, from https://www.zhycasting.com/determination-of-pouring-size-of-gating-system-in-lost-foam-casting-process-of-box/

Chapter 10
Die-Casting Processes

Nomenclature

t_s	Solidification time
T_{p-h}	Pre-heat temperature
$h_{init-gap}$	Initial heat transfer coefficient as a function of the gap width
T_{oil}	Oil tempering temperature
ϕ	Wedge angle
F_c	Clamping force
F_s	Separating force
p_i	Injection pressure
v	Plunger velocity
H	Plunger diameter
h	Depth of the molten metal in the sleeve
g	Gravitational acceleration
C	Cost per piece (total cost per piece)
C_M	Net material cost per piece
C_L	Labor cost per piece
C_D	Cost of die and tooling per piece
P	Production volume

Abbreviations

GDC	Gravity die casting
HPDC	High-pressure die casting
TPA	Total projection area

199

Z. Huda, *Metal Casting Engineering*, Mechanical Engineering Series, https://doi.org/10.1007/978-3-031-84620-5_10

10.1 Permanent-Mold Casting Processes

The permanent-mold casting processes are the metal casting processes that are based on permanent, multiple-use molds or dies. The main similarity of this group of casting processes lies in the employment of a permanent mold that can be used repeatedly for multiple metal castings. The mold, also called a die, is commonly made of alloy steel; however, other metals or ceramics can be used. The examples of components that may be manufactured in industry using a permanent-mold casting process include: cylinder blocks, cylinder heads, pistons, connecting rods, pipes, tubes, wheels, and parts for aircraft and rockets, gear blanks, and the like.

Broadly speaking, there are two types of permanent-mold casting processes: (1) die-casting processes and (2) centrifugal casting processes (see Chap. 1, Fig. 1.4). The die-casting processes can be classified into three types: (a) basic permanent-mold casting or gravity die casting (GDC), (b) low-pressure die casting, and (c) high-pressure die casting. These casting processes are described in the following sections.

10.2 Advantages and Limitations of Die Casting

10.2.1 Advantages of Die-Casting Processes

Permanent-mold/die-casting processes have several advantages as compared to expendable-mold casting; these advantages are listed below.

1. Permanent molds are generally made of alloy steel, which is a material with a high thermal conductivity. This thermal property of the mold material results in rapid cooling of the casting metal. This will result in a fine-grained microstructure, producing a casting with superior strength.
2. Uniform properties throughout the material of the cast part may be achieved with permanent-mold casting.
3. Another advantage of permanent-mold casting (over most expendable metal casting processes) is the achievement of closer dimensional accuracy as well as excellent surface finish of the part.
4. In industrial practice, permanent-mold casting results in a lower percentage of rejections than many expendable-mold processes.
5. The permanent-mold casting process can be highly automated.
6. This manufacturing process is useful in industry for mass production. When set up, it can be extremely economical with a high volume and rate of production.

10.2.2   Limitations of Die-Casting Processes

The die-casting process is only suited for materials with lower melting temperatures, such as zinc, copper, magnesium, and aluminum alloys. Although cast iron parts can be manufactured by this process, the high melting temperature of cast iron is hard on the mold. Steels are not generally cast by GDC because the high-temperature molten steel will cause damage (softening or deformation) of the die (molds). There is also a limitation to the size of cast parts manufactured by this process. The initial set-up costs are high, thereby making permanent-mold casting unsuitable for small production runs.

10.3   Basic Permanent-Mold Casting

10.3.1   What Is Basic Permanent-Mold Casting?

The basic permanent-mold casting, also called *gravity die casting (GDC)* involves the creation of a permanent mold. The sections of the mold are generally precisely machined from two separate metal blocks. These metal blocks (mold halves) are created so that they fit together and may be opened and closed easily and accurately (Fig. 10.1a, b). Additionally, the gating system as well as the part geometry is machined into the casting mold. The mold halves (when opened) are treated with a heat-resistant coating to endure the thermal cycles involved in the casting process. Then the mold halves are closed by use of a mold-clamp, which maintains the integrity of the mold by keeping the two halves securely together. Molten metal is then

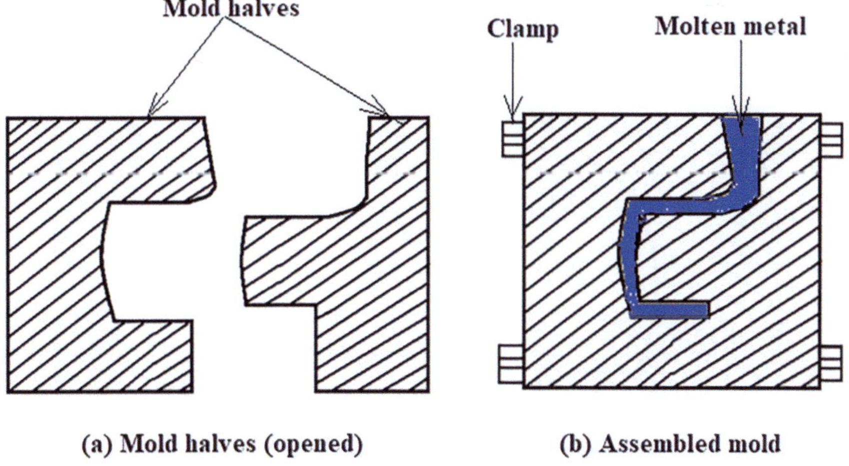

Fig. 10.1 Gravity die casting: (a) mold halves are open and (b) assembled mold

poured directly (by gravity) into the mold. Once the metal has solidified and cooled, the casting can be safely extracted from the mold.

The following are the principal steps in GDC: (1) mold creation, (2) treating the opened-molds with coating, (3) closing the molds and its pre-heating, (4) pouring the molten metal by gravity, and (5) ejecting the solidified and cooled casting. These operational steps are briefly explained in the following subsections.

10.3.2   Molds' Creation in GDC

The molds, in GDC, are generally made of high-strength alloy steels. A significant amount of resources is utilized in the production of the mold, making the set-up more expensive. However, once created, a permanent mold may be used thousands of times before its mold life expires. The number of castings that can be produced by a particular mold before it has to be replaced is termed *mold life*. The factors affecting mold life include: mold's operating temperature, mold material, casting metal, and the like.

10.3.3   Molds' Coating in GDC

Prior to pouring the molten for GDC, the internal surfaces of the permanent mold are coated by spraying with a slurry, which consists of refractory materials suspended in liquid (see Fig. 10.2). This coating tends to serve as a thermal insulator,

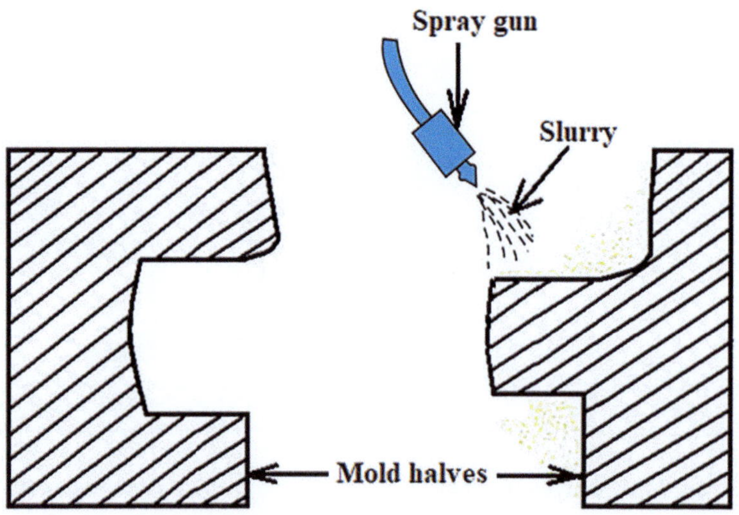

Fig. 10.2 Spraying (coating) the molds with slurry

thereby helping to control the heat flow and acting as a lubricant for easier removal of the cast part. Additionally, the application of the refractory coat as a regular part of the manufacturing process will increase the mold life of the valuable mold.

10.3.4   Closing the Mold and Its Pre-heating in GDC

Once the mold halves have been coated, the halves of the mold are closed and held together with the use of a clamp or some sort of mechanical means. In general, the mold is heated prior to the pouring of the metal casting. A possible temperature that a permanent mold may be heated to before pouring could be around 175 °C (350 °F). The pre-heating of the mold will facilitate the smoother flow of the molten metal through the mold's gating system and mold cavity. Pouring in a pre-heated mold will also reduce the thermal shock encountered by the mold due to the high-temperature gradient between the molten metal and the mold. It means that pre-heating of molds will act to increase mold life.

10.3.5   Pouring and Solidification

Once the molds have been securely closed and pre-heated, the molten metal is poured by gravity into the mold cavity for gravity die casting (GDC) (see Fig. 10.3). After pouring, the metal casting solidifies within the mold (Fig. 10.4).

Fig. 10.3 Pouring of molten metal into the permanent mold

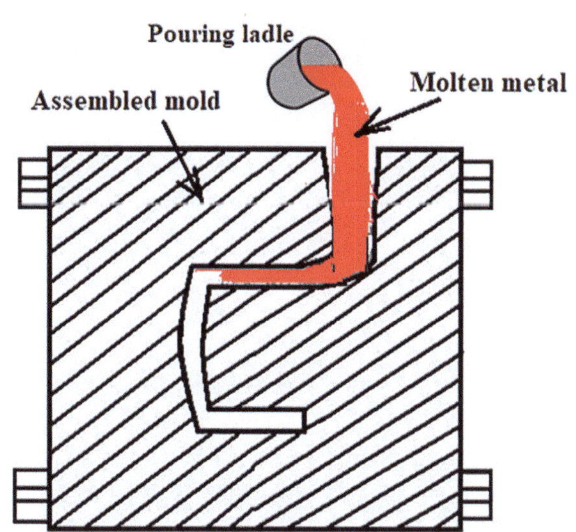

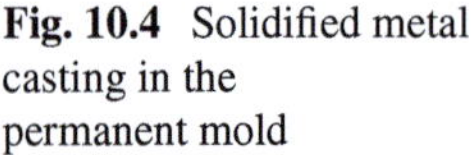

Fig. 10.4 Solidified metal casting in the permanent mold

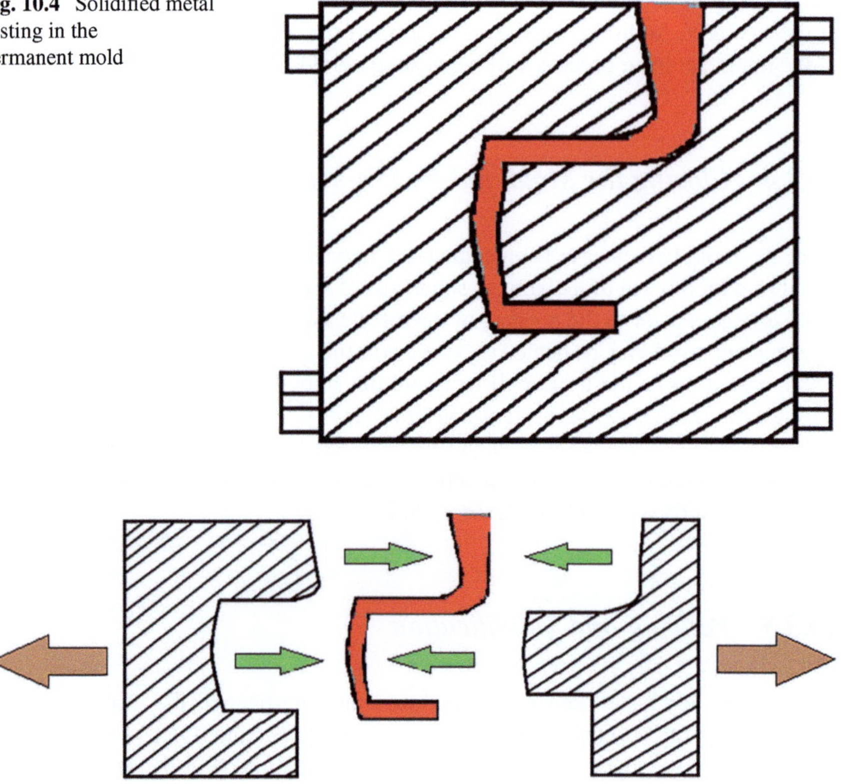

Fig. 10.5 Ejecting the casting in GDC

10.3.6   Ejecting the Casting

In the basic permanent-mold casting practice, the metal cast part is usually removed before the completion of cooling; this practice is done to prevent the solidified casting from contracting too much in the mold (see Fig. 10.5). This early ejection of casting prevents cracking of the casting, since the permanent mold does not collapse. The ejection of the casting is accomplished by means of ejector pins that are built into the mold.

10.3.7   Engineering Analysis of GDC

The mold pre-heating (see Sect. 10.3.4) decreases the temperature gradient within the casting. This results in a reduction in the solidification time of the casting. The solidification time can be related to the mold pre-heat temperature in the GDC of *Al-Si-Mg* alloys by (Reddy & Rajanna, 2009):

$$t_{s} = -0.2 \times 10^{-6} * T_{\mathrm{p\,h}}^{2} = 0.0002 * T_{\mathrm{p\,h}} + 9.9 \qquad (10.1)$$

where t_{s} is the solidification time (min), and $T_{\mathrm{p-h}}$ is the pre-heat temperature (°C) (see Example 10.1).

The thermal conditions strongly depend on the heat transfer between the casting and its surroundings. Local heat transfer coefficients describe how well heat can be transferred from one body or material to another. Recently (2022), Vossel and co-researchers reported the estimation of heat transfer coefficients in a gravity die casting (GDC) process with local air gap formation and heat shrinkage induced contact pressure. They investigated the casting of aluminum casting alloy A356. It has been experimentally found that the heat transfer coefficient is a function of the contact pressure and the gap width as well as the oil tempering temperature. The initial heat transfer coefficient as a function of the gap width can be determined by (Vossel et al., 2021):

$$h_{\mathrm{init\,gap}} = 0.00642 * T_{\mathrm{oil}}^{2} - 5.07 * T_{\mathrm{oil}} + 2229.4 \qquad (10.2)$$

where $h_{\mathrm{init-gap}}$ is the initial heat transfer coefficient as a function of the gap width (W/m²•K), and T_{oil} is the oil tempering temperature (K) (see Example 10.2).

10.4   Low-Pressure Die Casting

The low-pressure permanent-mold casting process involves the application of a controlled pressure in the range of 0.02–0.10 MPa (3–15 psi). This pressure is applied to a reservoir of molten metal to fill the permanent mold; the reservoir of molten metal is positioned below the mold (see Fig. 10.6). The gradual increments of gas pressure push the molten metal upwards through risers into the mold cavity.

The molten metal, in the reservoir, remains under pressure until it solidifies. Once the pressure has been released, any remaining molten metal in the risers and mold returns to the reservoir. The gradual application of pressure allows precise control over the mold filling process, ensuring a smooth and uniform distribution of metal. This technique helps to minimize porosity, reduces oxide formation, and ensures consistent metal flow throughout the mold. As a result the parts, produced by low-pressure die-casting process, have better density, superior strength, and precise dimensional accuracy.

10.5   High-Pressure Die Casting

10.5.1   The Dies in High-Pressure Die Casting

The high-pressure die-casting (HPDC) process involves forcing molten metal under high pressure (in the range of 1500–25,500 psi) into a die cavity, which is created using two high-strength alloy-steel dies that have been precisely machined into

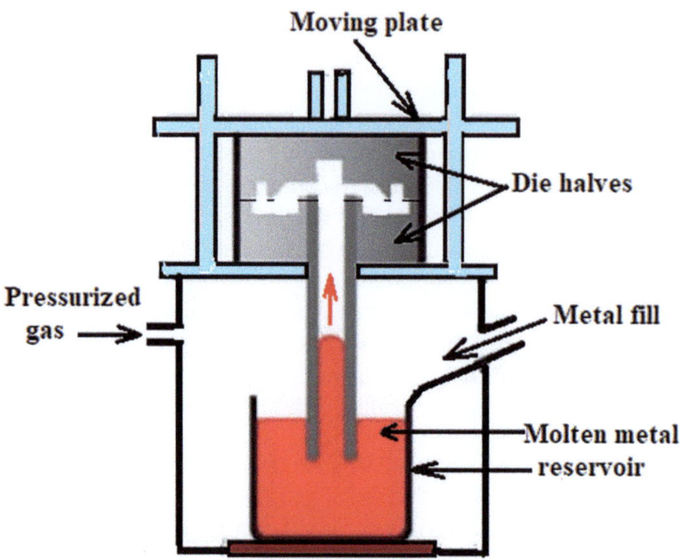

Fig. 10.6 Low-pressure die casting

shape. Depending on the type of metal being cast, a hot-chamber or cold-chamber die-casting machine is used. In both types of processes, the dies are cooled by spraying pressurized water. The dies, in the HPDC, are made of tool steel, die steel, or maraging steel (carbon-free iron-nickel alloys with additions of cobalt, molybdenum, titanium, and aluminum). The addition of tungsten and molybdenum imparts good refractory abilities and toughness to die-casting steel, thereby enabling the steel dies to withstand high temperature and pressure during the HPDC process.

10.5.2   Advantages and Limitations of HPDC

The HPDC process is highly economical for mass production of high-quality components, since this process is capable of producing an average of 500 castings per hour. Owing to the application of high pressure, the dimensional accuracy and surface finish are excellent (Andresen, 2004). This process is capable of casting components with thin sections. As the molten metal is in contact with the cold surface of the dies, the chilling effect results in small grain size in the material's microstructure. The fine-grains impart good strength to the casting.

There are a few limitations of HPDC. This process is generally limited to cast metals with low metal points, such as zinc, magnesium, lead, tin, etc. Additionally, the part geometry limits easy removal from the die cavity.

Table 10.1 Hot-chamber and cold-chamber HPDC processes

No.	Hot-chamber HPDC	Cold-chamber HPDC
1.	Crucible containing molten metal is placed in the furnace	The ladle containing molten metal is not immersed in the furnace
2.	Limited to process low-melting point metals (e.g. zinc, lead, tin, etc.)	This technology can process aluminum alloys
3.	Pressure range = 1000–5000 psi	Pressure range = 2000–20,000 psi
4.	The machine has a gooseneck	There is no gooseneck in this machine

10.5.3 Hot-Chamber and Cold-Chamber HPDC

Hot-chamber HPDC involves the use of a machine, in which the metal is contained in an open holding pot, placed in the furnace, and melted to the desired temperature. Conversely, cold-chamber HPDC uses a machine in which the shot chamber and plunger are not immersed in the molten metal, rather the molten metal is poured into the shot chamber. The difference between the hot-chamber and cold-chamber HPDC technologies is presented in Table 10.1.

10.5.4 Cold-Chamber HPDC Process

Cold-chamber die casting is a kind of high-pressure die casting in which the injection (shot) chamber and plunger are not immersed in the molten metal, but the pre-calculated molten metal is poured into the shot chamber (see Fig. 10.7). Once the molten metal has been poured into the shot chamber, the plunger pushes the molten metal into the die cavity at pressures in the range of 2000–20,000 psi (14–138 MPa). On the completion of solidification and cooling, the cast part is ejected by means of ejector pins. Cold-chamber die casting is the preferred HPDC process for metals having higher melting points. This is why cold-chamber die casting is mainly used for the production of aluminum alloy parts.

10.5.5 Hot-Chamber HPDC Process

The hot-chamber die casting is a HPDC process; it is so named because it involves heating of metal inside the casting machine. The application of hot-chamber die casting is limited to low melting point metals (e.g. zinc, tin, lead, magnesium, etc.) because these metals do not undergo hot corrosion when in contact with plungers and other mechanical components. In this process, metal is melted in a container (hot chamber); then a piston (plunger) forces molten metal under high pressure, in the range of 7–35 MPa (1000–5000 psi), into the die cavity. The hot-chamber

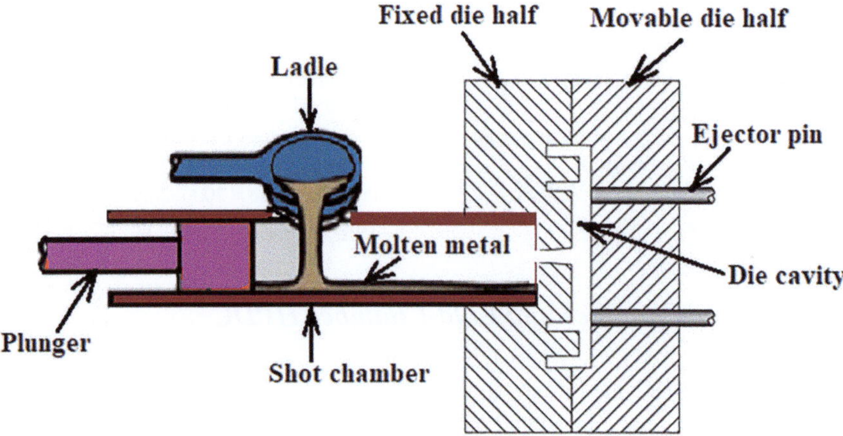

Fig. 10.7 Cold-chamber die casting

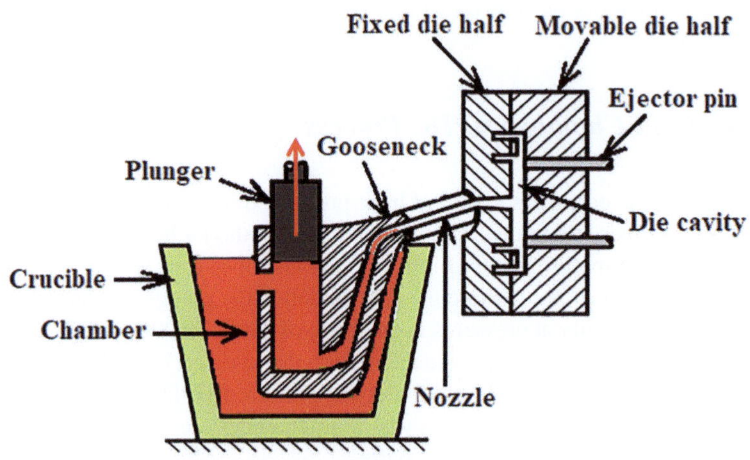

Fig. 10.8 Hot-chamber die-casting process with plunger withdrawn

die-casting process is well suited for high production rates, enabling us to cast 500 parts per hour.

The following are the four principal steps in the hot-chamber die-casting process: (1) lubricant coating, (2) with dies closed, the plunger is withdrawn, (3) plunger forces molten metal into the die cavity, and (4) the part is ejected. First, the dies are opened and the inside surfaces of the die cavity are coated with a lubricant (graphite) to facilitate easy removal of the casting. Metal is melted in the crucible (pot). Then the dies are closed, and the plunger is withdrawn, thereby allowing flow of molten metal from the crucible to the hot chamber; the metal also rises in the gooseneck (see Fig. 10.8). In the next step, the plunger exerts high pressure, thereby forcing metal to flow from the gooseneck via nozzle to fill the die cavity (see

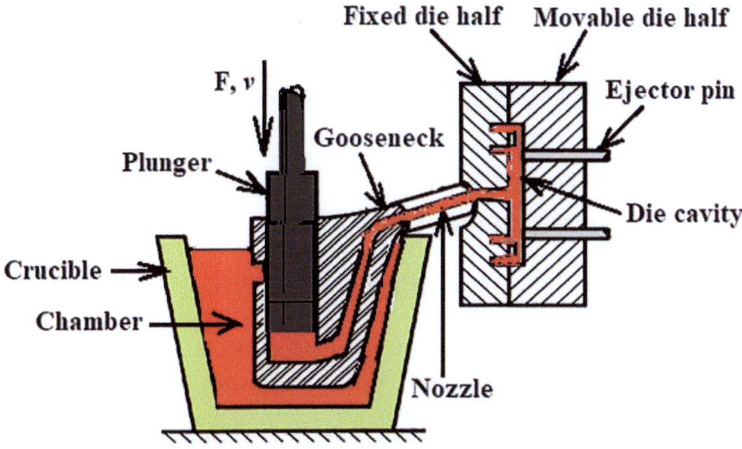

Fig. 10.9 High-pressure applied to force metal to fill the cavity in hot-chamber die casting

Fig. 10.9). The high pressure is maintained during cooling and solidification. Finally, the part (casting) is ejected.

10.5.6 Components of the Mold in a HPDC Machine

The mold of a HPDC machine is mainly composed of a fixed mold half and a movable mold half (see Fig. 10.9). The components of these mold halves include (a) core-pulling mechanism, (b) slider, (c) inclined-guide column, (d) wedge block, (e) limit block, (f) screws, springs, nuts, bolts, and the like. The *fixed mold half* is fixed on the mold mounting plate of the die-casting machine; it has a sprue connected to the nozzle or pressure chamber. The fixed mold half mainly consists of a fixed mold insert, fixed mold sleeve plate, guide pillar, wedge block, inclined-guide pillar, spruez sleeve, fixed mold core-pulling mechanism, and other parts. The *movable mold half* is fitted on the die-casting movable mold mounting plate and moves with the movable mold mounting plate to open and close the mold. The movable mold half is closed with the casting system to form a (die) cavity.

The *core-pulling mechanism* is used to extract parts with complex shapes (with holes) or grooves. It includes inclined-guide posts, side cores, sliders, guide slides, limit blocks, screws, springs, nuts and screws, and other components. The *slider* cooperates with the core and the inclined-guide column to complete the core-pulling action. The inclined-guide pillar facilitates the removal of the core from the casting. During the mold opening process, the slider is forced to pull out the core through the movement of the inclined-guide pillar. The *limit block* keeps the slider in a specific position to accurately insert the inclined-guide post when closing the mold. The *wedge block* withstands the counter pressure and prevents the slider from retreating during injection of the molten metal into the die cavity. The angle of wedge block or the *wedge angle* is generally related to the inclined-guide column angle by:

$$\text{Wedge angle} = \phi = \text{Inclined guide column angle} + 4° \qquad (10.3)$$

10.5.7 Engineering Analysis of HPDC

10.5.7.1 Determination of Clamping Force in HPDC

The die-casting clamping force is the force applied to a mold by the clamping unit of HPDC machine. The clamping force F_c must be greater than the separating force F_s, according to the following relationship:

$$F_c = 1.2 * F_s \qquad (10.4)$$

The separation force can be determined by:

$$F_s = \text{TPA} * p_i \qquad (10.5)$$

where F_s is the separation force (N), TPA is the total projection area (cm^2), and p_i is the injection pressure (kgf/cm^2). In order to determine the separating force F_s, it is necessary to first calculate the total projection area, taking into consideration the casting, the slide, the runner, and the overflow projection areas, as follows:

$$\text{TPA} = \text{Casting projection area} + \text{slide p.a} + \text{runner p.a} + \text{overflow p.a} \quad (10.6)$$

Once the casting projection area has been calculated, the sum of runner and overflow projection areas can be determined by:

$$\left(\text{Runner p.a} + \text{overflow p.a}\right) = 40\% * \left(\text{casting projection area}\right) \qquad (10.7)$$

The slide core projection area can be determined by:

$$\text{Slide core p.a} = \left(\text{p.a in the plane normal to the core travel direction}\right) * \tan\phi \qquad (10.8)$$

where ϕ is the wedge angle (deg).
The significance of Eqs. 10.3–10.8 is illustrated in Examples 10.3–10.8.

10.5.7.2 Determination of Plunger Velocity in a HPDC

In order to exert the required pressure on molten metal for injection to the die cavity, the plunger must move with correct velocity. The plunger velocity can be determined by (Reikher & Barkhudarov, 2007):

$$v = 2 * \left(\sqrt{g*H} - \sqrt{g*h} \right) \tag{10.9}$$

where v is the plunger velocity (m/s), H is the plunger diameter (m), h is the depth of the molten metal in the sleeve (m), and g is the gravitational acceleration (see Example 10.9). It has been experimentally shown that when the specified (defined) plunger velocity is too slow, excessive metal splashing results in the shot chamber of the die-casting machine (see Fig. 10.7). Conversely, when the specified plunger velocity is too high, wave velocity excites the critical wave form, resulting in air entrainment defect. Based on several velocities tried, the final slow shot velocity was found to be 0.3 m/s (Reikher & Gerrber, 2011).

10.5.7.3 Casting Cost in HPDC

The cost of casting in a HPDC (taking into account only the casting process) can be reduced by increasing the number of pressure die cavities. This cost reduction often requires changing the die-casting machine to a machine with a higher clamping force, and due to the larger volume of the die cavity, it is necessary to use vacuum systems (Dudek et al., 2023). The casting cost can be calculated by using the following simplified formula (Lenau & Egebøl, 1995):

$$\text{Casting cost} = \frac{\text{Cycle time}}{\text{No.of cavities}} * \text{MHR} \tag{10.10}$$

where MHR is the machine hourly rate (\$/h). The MHR can be calculated by:

$$\text{MHR} = \frac{\text{Manufacturing overheads}}{\text{Machine hours}} \tag{10.11}$$

The machine cycle time can be determined by:

$$\text{Cycle time} = \frac{\text{Machine hours}}{\text{No.of castings produced}} \tag{10.12}$$

The significance of Eqs. 10.10–10.12 is illustrated in Examples 10.10–10.13.

10.5.7.4 Determination of the Cost of HPDC Product

The total cost of a product (piece), produced by HPDC, can be determined by:

$$C = C_{\text{M}} + C_{\text{L}} + \frac{C_{\text{D}}}{P} \tag{10.13}$$

where C is the total cost per piece (\$/piece), C_M is the net material cost per piece, C_L is the labor cost per piece, C_D is the cost of die and tooling per piece, and P is the production volume (number of pieces produced).

The cost of material is not only the cost of purchasing the alloy but also the costs associated with the melting process (energy, non-returnable losses, etc.). This means that we need to use more material to complete the die casting. This involves the melting loss of the material, the alloying elements added, and the material of the gate. By assuming that a unit piece (casting) weighing 1 kgf, the material cost can be expressed by the following relationship:

$$C_M = \text{Materials}(\text{purchasing})\,\text{cost} + \text{material loss costs} \qquad (10.14)$$

The material loss cost is usually 1–8% of the material cost. The labor cost includes the cost incurred on de-burring, quality inspection, packaging, etc., which account for about 2% of the cost of material. The cost of die and tooling involves the costs of die material, design, processing, etc. The significance of Eqs. 10.13–10.14 is illustrated in Examples 10.14 and 10.15.

10.6   Worked Numerical Examples in Die Casting

Example 10.1 Determining the Solidification Time in GDC

An *Al-Si-Mg* alloy component was cast by GDC. The pre-heat temperature of the permanent mold is 200 °C. Determine the solidification time.

Solution

By using Eq. 10.1,

$$t_s = -0.2 \times 10^{-6} * T_{p\text{-}h}^2 - 0.0002 * T_{p\text{-}h} + 9.9 = -0.2 \times 10^{-6} \times 200^2 - 0.0002 \times 200 + 9.9$$

$$t_s = -0.008 - 0.04 + 9.9 = 9.85\,\text{min}$$

The solidification time = 9.85 min.

Example 10.2 Determination of the Initial Heat Transfer Coefficient in GDC

An aluminum alloy A356 component is to be cast by GDC with local air gap formation and contact pressure. The oil tempering temperature is 100 °C. Determine the initial heat transfer coefficient as a function of the gap width.

Solution

$$T_{oil} = 100°C = 100 + 273 = 373\,\text{K}$$

By using Eq. 10.2,

$$h_{\text{init gap}} = 0.00642 * T_{\text{oil}}^2 - 5.07 * T_{\text{oil}} + 2229.4 = 0.00642 \times 373^2 - 5.07 \times 373 + 2229.4$$

$$h_{\text{init-gap}} = 893.21 - 1891 + 2229.4 = 1231.6 \, \text{W} / \text{m}^2 \cdot \text{K}$$

The initial heat transfer coefficient as a function of the gap width = 1231.6 W/m²·K

Example 10.3 Determination of the Sum of Runner and Overflow Projection Areas

An aluminum component casting (with a projection area of 550 cm²) is to be produced in HPDC machine with an inclined-guide column angle of 17°. The projection area, in the plane normal to the core travel direction, is 55 cm². The injection pressure for the casting is 700 kgf/cm². Determine the sum of the runner and overflow projection areas.

Solution

Casting projection area = 550 cm², Runner p.a + overflow p.a =?
By using Eq. 10.7,

$$\text{Runner p.a} + \text{overflow p.a} = 40\% * \left(\text{casting projection area}\right) = 0.4 \times 550 = 220 \, \text{cm}^2$$

Runner projection area + overflow projection area = 220 cm².

Example 10.4 Determining the Wedge Block Angle for HPDC Operation

By using the data in Example 10.3, determine the angle of wedge block in the HPDC machine.

Solution

Inclined-guide column angle = 17°, Wedge angle = ϕ =?
By using Eq. 10.3,

$$\text{Wedge angle} = \phi = \text{Inclined guide column angle} + 4° = 17 + 4 = 21°$$

Example 10.5 Determining the Slide Core Projection Area for HPDC Operation

By using the data in Examples 10.3 and 10.4, determine the slide core projection area.

Solution

The projection area, in the plane normal to the core travel direction, = 55 cm², $\phi = 21°$.
By using Eq. 10.8,

$$\text{Slide core p.a} = \left(\text{p.a in the plane normal to the core travel direction}\right) * \tan \phi$$
$$= 55 \times \tan 21°$$

$$\text{Slide core projection area} = 55 \times 0.3838 = 21.11 \, \text{cm}^2$$

Example 10.6 Determining the Total Projection Area for HPDC Operation
By using the data in Examples 10.3–10.5, determine the total projection area for the
HPDC operation.

Solution
Casting projection area = 550 cm^2, Slide core projection area = 21.11 cm^2.

Runner projection area + overflow projection area = 220 cm^2, Total projection
area = TPA =?

By using Eq. 10.6.

$$TPA = \text{Casting p.a} + \text{slide p.a} + \text{runner p.a} + \text{overflow p.a}$$
$$= 550 + 21.11 + 220 = 791.11 \, \text{cm}^2$$

Total projection area = 791.11 cm^2.

Example 10.7 Determining the Separating Force for HPDC Operation
By using the data in Examples 10.3–10.6, determine the separating force in the
HPDC operation.

Solution
Total projection area = TPA = 791.11 cm^2, p_i = 700 kgf/cm^2, F_s =?

By using Eq. 10.5,

$$F_s = TPA * p_i = 791.11 \times 700 = 553,777 \, \text{kgf}$$

The separating force = F_s = 553,777 kgf.

**Example 10.8 Determining the Clamping Force and Selecting the Tonnage
of Machine**
By using the data in Example 10.7, determine the clamping force for the HPDC
operation. Select also the tonnage of HPDC machine to be used.

Solution
By using Eq. 10.4,

$$F_c = 1.2 * F_s = 1.2 \times 553,777 = 6,645,324 \, \text{kgf} = 664.532 \, \text{ton}$$

The clamping force = 664.532 ton.

A 700-ton HPDC machine may be selected. It would be better to select 800-ton
machine.

Example 10.9 Designing a Plunger for HPDC Machine
The optimum slow shot velocity of the plunger for a HPDC machine is 0.3 m/s. The
die-casting process requires the depth of the molten metal in the shot sleeve to be
5 cm. Design the plunger by specifying its diameter.

Solution
v = 0.3 m/s, h = 5 cm = 0.05 m, g = 9.81 m/s^2, H =?

By using Eq. 10.9,

$$0.3 = 2 \times \left(\sqrt{9.81 \times H} - \sqrt{9.81 \times 0.05} \right)$$

$$0.3 = 2 \times \left(\sqrt{9.81 \times H} - 0.7 \right)$$

$$\sqrt{9.81 \times H} - 0.7 = 0.15$$

$$\sqrt{9.81 \times H} = 0.85$$

$$9.81 \times H = 0.7225$$

$$H = 0.0736\,\mathrm{m} = 7.36\,\mathrm{cm}$$

The plunger diameter $\cong$ 8 cm.

Example 10.10 Determining the Machine Hourly Rate in HPDC

The manufacturing overheads in the operation of a HPDC machine are $1000 when the machine is used for 80 h. There are five die cavities in the machine. The die-casting machine produces 400 pieces per hour. Determine the machine hourly rate for the HPDC operation.

Solution

Manufacturing overheads = $1000, Machine hours = 80, MHR =?
  By using Eq. 10.11,

$$\mathrm{MHR} = \frac{\text{Manufacturing overheads}}{\text{Machine hours}} = \frac{\$1000}{80} = \$12.5 / \mathrm{h}$$

Machine hourly rate = MHR = $12.5/h.

Example 10.11 Determining the Cycle Time in HPDC

By using the data in Example 10.10, determine the cycle time for the HPDC operation.

Solution

Machine hours = 80 h, No. of casting produced = 400 × 80 = 32,000, Cycle time −?
  By using Eq. 10.12,

$$\text{Cycle time} = \frac{\text{Machine hours}}{\text{No.of castings produced}} = \frac{80\,\mathrm{h}}{32,000\,\mathrm{pcs}} = 0.0025\,\mathrm{h} / \text{piece}$$

The cycle time = 0.0025 h/casting.

Example 10.12 Determining the Casting Cost in HPDC

By using the data in Examples 10.10 and 10.11, determine the casting cost for the HPDC operation.

Solution

MHR = \$7.5/h, Cycle time = 0.0025 h/piece, No. of die cavities = 5, Casting cost =?
   By using Eq. 10.10,

$$\text{Casting cost} = \frac{\text{Cycle time}}{\text{No.of cavities}} * \text{MHR} = \frac{0.0025 \ \text{h / piece}}{5} \times \$12.5 / \text{h}$$
$$= \$0.00625 / \text{piece}$$

Casting cost $\cong$ \$ 0.00625/piece.

Example 10.13 Determining the Net Material Cost in HPDC

It is required to produce an aluminum casting by HPDC process. The component weighs 1 kg, and its material (purchase) cost is USD 2.60. The material loss cost is 7% of the material purchase cost. The labor cost is 1.5% of the material cost. The cost of dies and tooling is USD 1000.00. The lifespan of the die and tooling of HPDC machine is designed to produce 5000 castings. Determine the net material cost.

Solution

Material purchase cost = USD 2.6, Material cost =?, Labor cost =?

$$\text{Material loss cost} = 7\% \text{of } 2.6 = \frac{7}{100} \times 2.6 = \text{USD} \, 0.18$$

By using Eq. 10.14,

$$C_M = \text{Materials} \left(\text{purchasing} \right) \text{cost} + \text{material loss costs}$$

$$C_M = 2.6 + 0.18 = \text{USD} \, 2.78 / \text{piece}$$

Net material cost = USD 2.78/piece.

Example 10.14 Determining the Labor Cost in HPDC

By using the data in Example 10.13, determine the labor cost.

Solution

Labor cost is 2% of the net material cost.

$$C_L = \frac{2}{100} \times 2.78 = \text{USD} \, 0.055 / \text{piece}$$

Labor cost = USD 0.055/piece.

Example 10.15 Determining the Total Cost of a Die-casting Product

By using the data in Examples 10.13 and 10.14, determine the total cost of a piece (product).

Solution

C_M = USD 2.78/piece, C_L = USD 0.055/piece, C_D = USD 1000.00, P = 5000 pieces.
 By using Eq. 10.13,

$$C = C_M + C_L + \frac{C_D}{P} = 2.78 + 0.055 + \frac{1000}{5000} = USD\,3\,/\,piece$$

The total cost of a casting (product) = USD 3.00.

Questions and Problems

10.1. Encircle the most appropriate answers for each of the following statements.

   (1) Which metal is used for making dies for die casting?
        (a) Aluminum alloy, (b) lead, (c) alloy steel, (d) brass
   (2) Which casting process involves the pouring of metal into mold by gravity?
        (a) Basic permanent-mold casting, (b) low-pressure die casting, (c) HPDC
   (3) What is the range of pressure in low-pressure die casting?
        (a) 0.5–1.0 psi, (b) 3–15 psi, (c) 50–100 psi, (d) 0.001–0.1 psi
   (4) Which type of die casting is most extensively used for producing complex-shaped parts?
        (a) Basic permanent-mold casting, (b) low-pressure die casting, (c) HPDC
   (5) Which one is the lowest-cost equipment cost casting process?
        (a) Basic permanent-mold casting, (b) low-pressure die casting, (c) HPDC
   (6) Which die casting machine has a gooseneck?
        (a) GDC, (b) low-pressure DC, (c) cold-chamber HPDC, (d) hot-chamber HPDC
   (7) Which die-casting machine is mainly used for the production of aluminum alloy parts?
        (a) GDC, (b) cold-chamber HPDC, (c) hot-chamber HPDC, (d) low-pressure DC
   (8) What range of pressure is used in high-pressure die-casting machine?
        (a) 2000–20,000 psi, (b) 1000–2000 psi, (c) 100–1000 psi, (d) 5–100 psi

10.2. Why is die casting unsuitable for low volume of production?
10.3. Why is die casting process unsuitable for producing steel parts?
10.4. Explain gravity die casting (GDC) process with reference to each step of operation.
10.5. Describe low-pressure die casting process with the aid of a sketch.
10.6. Differentiate between the cold-chamber die-casting and hot-chamber die casting machines.
10.7. Explain cold-chamber die-casting process with the aid of a sketch.

P-10.8. A metallic component (with a projection area of 500 cm^2) is to be cast in an HPDC machine with an inclined-guide column angle of 15°. The projection area, in the plane normal to the core travel direction, is 50 cm^2. The injection pressure for the casting is 720 kgf/cm^2. Determine the sum of the runner and overflow projection areas.

P-10.9. By using the data in P-10.8, determine the clamping force for the die casting operation.

P-10.10. The manufacturing overheads in the operation of a HPDC machine are $800 when the machine is used for 70 h. There are four die cavities in the machine. The die-casting machine produces 500 pieces per hour. Determine the machine hourly rate for the HPDC operation.

P-10.11. By using the data in P-10.10, determine the casting cost for the HPDC operation.

P-10.12. It is required to cast a zinc component by HPDC process. The component weighs 1 kg, and its material (purchase) cost is USD 3.00. The material loss cost is 8% of the material purchase cost. The labor cost is 1.5% of the material cost. The cost of dies and tooling is USD 1100.00. The lifespan of the die and tooling of HPDC machine is designed to produce 5000 castings. Determine the net material cost.

P-10.13. By using the data in P-10.12, determine the total cost of a piece (product).

References

Andresen, B. (2004). *Die cast engineering: A hydraulic, thermal, and mechanical process.* CRC Press.

Dudek, P., Białoń, J., Piwowońska, J., Walczak, W., & Wrzała, K. (2023). The impact on the cost of making high pressure die castings with multi-cavity die and vacuum assistance. *Vacuum, 210*, 111859.

Lenau, T., & Egebøl, T. (1995). Early cost estimation for die casting. In *Proceedings of ICED 1995*, Edition Heurista, Praha (pp. 1007–1016). http://www.polynet.dk/lenau/early.htm

Reddy, A. C., & Rajanna, C. (2009). Design of gravity die casting process parameters for *Al-Si-Mg* alloys. *Journal of Machining and Forming Technologies, 1*(1/2), 1–25. https://jntuhceh.ac.in/web/tutorials/faculty/779_I-07.pdf

Reikher, A., & Barkhudarov, M. R. (2007). *Casting: An analytical approach.* Springer.

Reikher, A., & Gerrber, H. (2011). Calculation of the die cast parameters of the thin wall aluminum die cast part. In *2011 Die Casting Congress and TableTop*. NADCA. https://www.flow3d.com/wp-content/uploads/2014/08/Calculation-of-the-Die-Cast-parameters-of-the-Thin-Wall-Aluminum-Cast-Part.pdf

Vossel, T., Wolff, N., Pustal, B., & hrig-Polaczek, A. Bu. (2021) Heat transfer coefficient determination in a gravity die casting process with local air gap formation and contact pressure using experiemntal evaluation and numerical simulation. International Journal of Metalcasting. https://doi.org/10.1007/s40962-021-00663-y

Chapter 11
Centrifugal Casting Processes

Nomenclature

F	Centrifugal force
v	Tangential velocity of the metal
R	Internal radius of the mold
G_f	G-factor = Gravity factor
N	Rotational speed of the mold
D	Inside diameter of mold = outside diameter of the casting
n_1	Rotational speed of the driving pulley
d_1	Diameter of the driving pulley
n_2	Rotational speed of the driven pulley
d_2	Diameter of the driven pulley
v_1	Peripheral velocity of the belt on the driving pulley
v_2	Peripheral velocity of the belt on the driven pulley
L	Total length of the belt on the two pulleys
c	Distance between the centers of the two pulleys

Abbreviations

SCC	Semi-centrifugal casting
TCC	True centrifugal casting

11.1 Working Principle of Centrifugal Casting

Centrifugal casting is so named because it uses rotation's centrifugal forces to distribute the molten metal to the outer regions of a cylinderical mold cavity, where it solidifies to create a part. In *centrifugal casting,* a permanent mold is rotated

© The Author(s), under exclusive license to Springer Nature
Switzerland AG 2025

Z. Huda, *Metal Casting Engineering*, Mechanical Engineering Series,
https://doi.org/10.1007/978-3-031-84620-5_11

continuously about its axis at high speeds (in the range of 300–3000 rpm) as the molten metal is poured into the mold. As the mold rotates, the molten metal is centrifugally thrown toward the inside mold wall, where it cools and solidifies. Owing to the chilling effect on the mold surface, the outer skin of the casting has a fine-grained microstructure. The impurities and non-metallic inclusions in the molten metal are also centrifugally thrown to the inside mold wall (outer skin of the casting). These impurities can be machined away by internal cylindrical grinding (Campbell, 2011). There are three forms of centrifugal casting: (a) true centrifugal casting, (b) semi-centrifugal casting, and (c) centrifuge casting. The three forms of centrifugal casting are discussed in the following sections.

11.2   Advantages and Limitations of Centrifugal Casting

The centrifugal casting is an ideal metal casting process to produce thin-walled cylindrical components, such as bushings, engine cylinder liners, rings, brake drums, water supply lines, sewage pipes, street-lamp posts, and gas pipes. These cylindrical components are used in high-reliability applications such as jet-engine compressor cases, hydro-wear rings, and numerous military items because they produce components with excellent material soundness due to very high centrifugal forces. The outer skin of the casting is fine-grained, thereby imparting high strength to the material. Centrifugal casting allows the processing of a wide range of materials, including iron, carbon steel, stainless steel, and alloys of aluminum, copper, and nickel. Additionally, composites can be cast, since two materials can be cast together by introducing a second material during the process.

The main limitation of this process is that only cylindrical shapes can be produced. In true centrifugal casting, the wall thickness of the cylinder must be in the range of 2.5–125 mm.

11.3   True Centrifugal Casting

True centrifugal casting (TCC) uses a hollow, cylindrical mold that is rotated about a vertical or horizontal axis. The mold is typically made of iron, steel, or graphite (see Fig. 11.1). The molds are made of steel, iron, or graphite. They can be coated with a refractory lining to extend their life. The surfaces of the molds are designed to allow for pipe casting with a range of outside styles. The inside surface of the casting remains cylindrical due to centrifugal forces that evenly distribute the molten metal. The outer surface of the mold is cooled by a water flowing system (Romanoff, 1981).

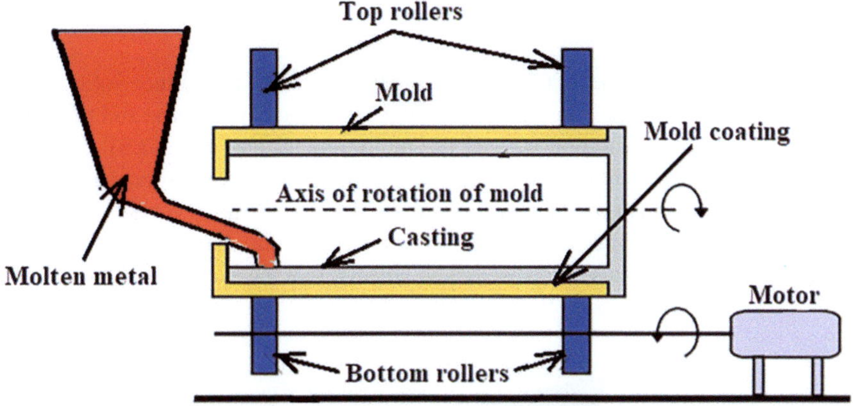

Fig. 11.1 True centrifugal casting (TCC) with horizontal axis of rotation

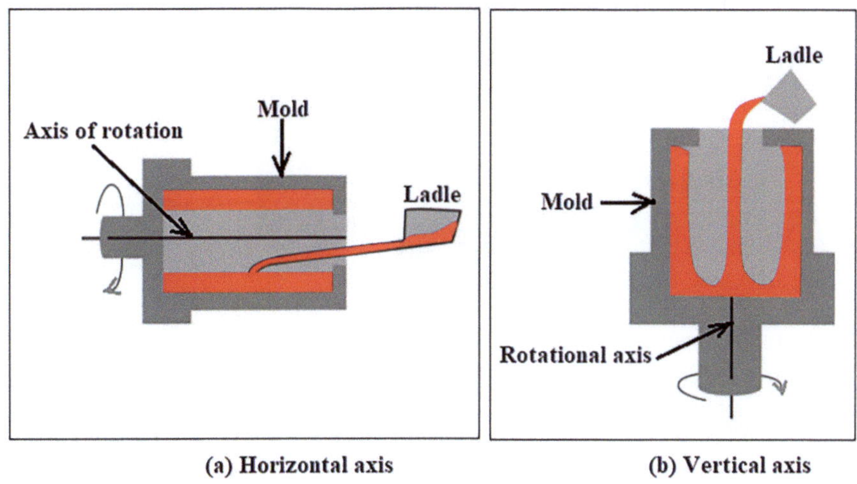

Fig. 11.2 Horizontal TCC (**a**) and vertical TCC (**b**)

There are two types of true centrifugal casting: (a) horizontal TCC and (b) vertical TCC. The horizontal TCC is performed on a horizontal axis of rotation of the mold, whereas the vertical TCC involves a vertical rotational axis (see Fig. 11.2a, b). The horizontal TCC is explained in the preceding paragraph. In vertical TCC, the effect of gravity acting on the molten metal causes the casting wall to be thicker at the base than at the top. The inside profile of the casting wall takes the shape of a parabola in vertical TCC.

11.4   Semi-centrifugal Casting

Semi-centrifugal casting (SCC) involves centrifugal force to produce solid casting rather than tubular shapes. In this process, the molds have their sprue and riser positioned at the axis of rotation to feed the molten metal (see Fig. 11.3). The outer regions of the parts, produced by SCC, have a higher density than the center of the rotating axis. This material behavior is due to higher centrifugal forces in the outer region during the casting process (see Fig. 11.4). The semi-centrifugal casting process can cast components (such as spoked wheels) that have a rotational symmetry.

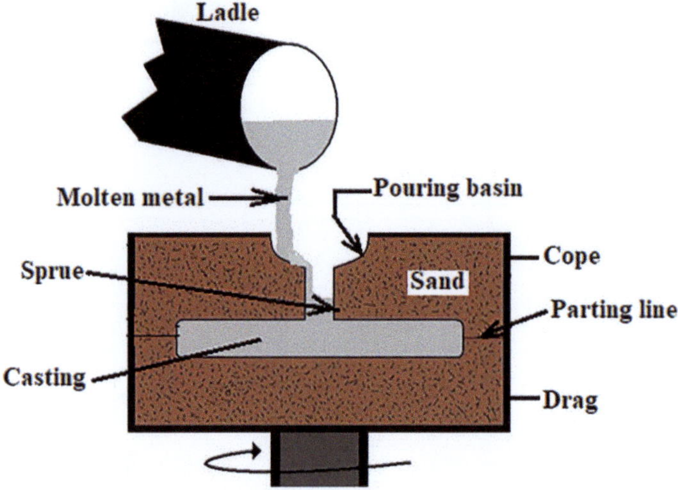

Fig. 11.3 Semi-centrifugal casting process

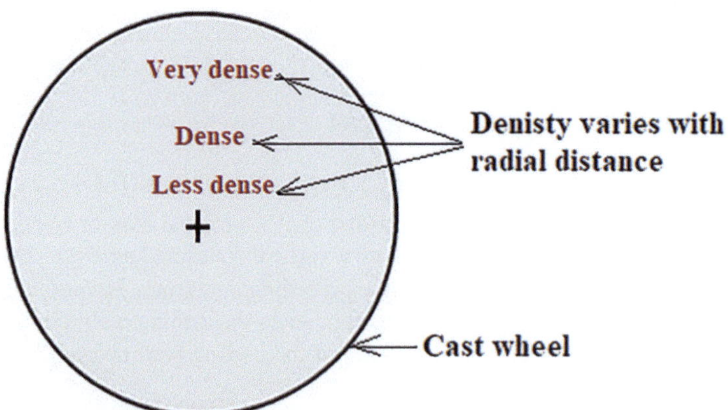

Fig. 11.4 Semi-centrifugally cast wheel

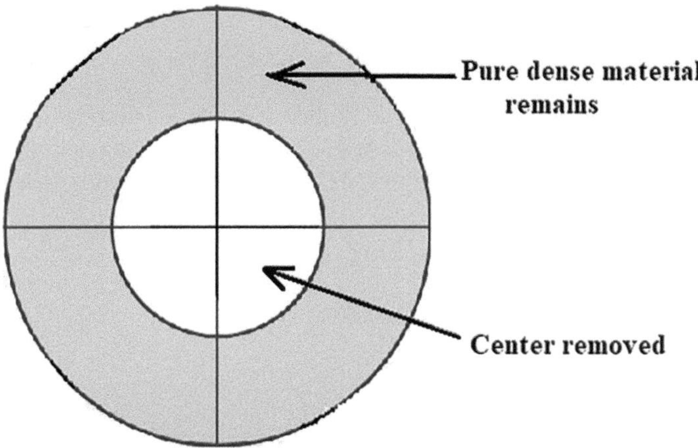

Fig. 11.5 Semi-centrifugally cast wheel with removed center

In semi-centrifugally cast wheels, the impurities and inclusions (with lesser density) are accumulated near the center of the wheel due to higher centripetal forces. These less-dense impurities are generally machined out from the center in industrial practice (see Fig. 11.5).

11.5   Centrifuge Casting

Centrifuge casting, also known as centrifuging, involves the placement of mold cavities of any shape at a certain distance from the axis of rotation. The molten metal is poured from the center, and centrifugal forces push the metal into the mold cavity through the sprue and the runner. Like true centrifugal casting, the characteristics of the castings might change with distance from the rotational axis. *Centrifuging* is used for casting of small parts (e.g. jewelry, small bushes, sleaves, and the like).

11.6   Engineering Analysis of True Centrifugal Casting

11.6.1   *Derivation of an Expression for the Mold Rotational Speed*

The centrifugal force acting on the molten metal, in true centrifugal casting, is given by:

$$F = \frac{m \cdot v^2}{R} \tag{11.1}$$

where F is the centrifugal force (N), m is the mass of molten metal (kg), v is the tangential velocity of the metal (m/s), and R is the internal radius of the mold (m). The ratio of the centrifugal force to the weight of the molten metal is called *G-factor* (G_f), which is expressed as:

$$G_f = \frac{F}{\text{Weight}} = \frac{F}{m \cdot g} \tag{11.2}$$

By combination of Eqs. 11.1 and 11.2, we obtain:

$$G_f = \frac{F}{m \cdot g} = \frac{\dfrac{m \cdot v^2}{R}}{m \cdot g} = \frac{v^2}{R \cdot g} \tag{11.3}$$

The tangential speed of the molten metal is related to the rotational speed of the mold by:

$$v = 2\pi \cdot R \cdot N \tag{11.4}$$

where v is the tangential speed (m/s), and N is the rotational speed (rev/s). Since rotational speed of the mold, N, is expressed in rev/min, Eq. 11.4 can be re-written as:

$$v = \frac{2\pi \cdot R \cdot N}{60} = \frac{\pi \cdot R \cdot N}{30} \tag{11.5}$$

where the rotational speed N is expressed in rev/min. The substitution of the value of v (from Eq. 11.5) into Eq. 11.2 yields:

$$G_f = \frac{v^2}{R \cdot g} = \frac{\left(\dfrac{\pi \cdot R \cdot N}{30}\right)^2}{R \cdot g} \tag{11.6}$$

By solving Eq. 11.6 for N, we obtain:

$$N = \frac{30}{\pi} * \sqrt{\frac{2 * G_f * g}{D}} \tag{11.7}$$

where N is the rotational speed (rev/min), G_f is the G-factor ($G = 60$–68), D is the inside diameter of mold (m) = outside diameter of the casting (m), $g = 9.8$ m/s^2.

The significance of Eqs. 11.1–11.7 is illustrated in Examples 11.1–11.7.

In TCC with vertical axis, the rotational speed of the mold is related to the inside radii at the top and bottom of the casting, according the following formula:

$$N = \frac{30}{\pi} * \sqrt{\frac{2*g*L}{\left(R_t^2 - R_b^2\right)}} \tag{11.8}$$

where L is the vertical length of the casting (m), R_t is the inside radius at the top of the casting, (m), and R_b is the inside radius at the bottom of the casting (m) (see Examples 11.8 and 11.9). The part length must not exceed two times the diameter of the part in this process.

11.6.2   Design of Transmission System from Motor to Centrifugal Casting Machine

The operating principle of centrifugal casting generally involves the rotation of a metallic mold by the use of a motor and a two-pulley-belt system (see Fig. 11.6). Additionally, an inverter is used to allow the regulation of the motor rotational speed. The driving pulley is linked to the motor, whereas the driven pulley is linked to the mold via appropriate mechanical mechanisms. The two-pulley-belt (open-belt drive) system can be designed using the following relationship:

$$\frac{n_1}{n_2} = \frac{d_2}{d_1} \tag{11.9}$$

where n_1 is the rotational speed of the driving pulley (RPM), d_1 is the diameter of the driving pulley, n_2 is the rotational speed of the driven pulley (RPM), and d_2 is the diameter of the driven pulley (see Example 11.10).

The peripheral speed of the belt on the driving pulley can be determined by (Khurmi & Gupta, 2005):

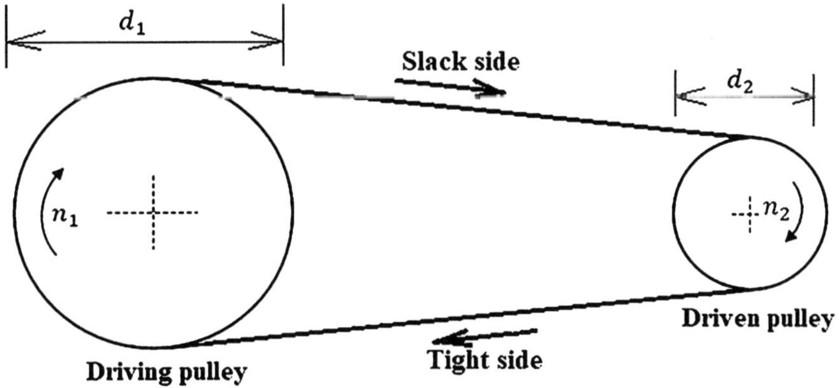

Fig. 11.6 Two-pulley-belt system (open-belt drive) for centrifugal casting machine

$$v_1 = \frac{\pi \cdot d_1 \cdot n_1}{60} \tag{11.10}$$

where v_1 is the peripheral velocity of the belt on the driving pulley (m/s), d_1 is the driving-pulley diameter (m), and n_1 is the rotational speed of the driving pulley (rev/min) (see Example 11.11).

Similarly, the peripheral speed of the belt on the driving pulley can be calculated by:

$$v_2 = \frac{\pi \cdot d_2 \cdot n_2}{60} \tag{11.11}$$

where v_2 is the peripheral velocity of the belt on the driven pulley (m/s), d_2 is the driven-pulley diameter (m), and n_2 is the rotational speed of the driven pulley (rpm) (see Examples 11.12 and 11.13).

The total length of the belt for operating the centrifugal casting machine can be determined by (Oyewole & Sunday, 2011):

$$L = 2 \cdot c + \frac{\pi}{2} \cdot \left(d_1 + d_2\right) + \frac{\left(d_1 - d_2\right)^2}{4 \cdot c} \tag{11.12}$$

where L is the total length of the belt (mm), and c is the distance between the centers of the two pulleys (see Example 11.14).

11.7   Worked Numerical Examples in Centrifugal Casting

Example 11.1 Determining the Mass of Metal To Be Poured in True Centrifugal Casting

A copper hollow cylinder with $OD = 10$ mm, $ID = 3$ mm, and $L = 150$ cm is required to be manufactured by true centrifugal casting. Calculate the mass of copper required to be poured for the casting process. Ignore all allowances. The density of copper is 8.96 g/cm³.

Solution

$L = 150$ cm, $OD = 10$ mm $= 1$ cm, $ID = 3$ mm $= 0.3$ cm, $\rho = 8.96$ g/cm³

$$V = \text{Volume of metal in the cylinder} = \text{Outer volume} - \text{Inner volume} = V_o - V_i$$

$$V = \frac{\pi}{4} * \left[(OD)^2 - (ID)^2\right] * L = 0.785 \times \left[(1)^2 - (0.3)^2\right] \times 150 = 107.15\,\text{cm}^3$$

$$\text{Mass} = \rho * V = 8.96 \times 107.15 = 960\,\text{g}$$

The mass of copper required to be poured = 960 g $\cong$ 1 kg.

Example 11.2 Determining the Rotational Speed of Mold in True Centrifugal Casting (TCC)

By using the data in Example 11.2, determine the rotational speed of the mold for the TCC, if G-factor is 64.

Solution

D = outside diameter of the casting = 1 cm = 0.01 m, G_f = 64, N =?

By using Eq. 11.7,

$$N = \frac{30}{\pi} * \sqrt{\frac{2 * G_f * g}{D}} = \frac{30}{\pi} \times \sqrt{\frac{2 \times 64 \times 9.81}{0.01}} = 9.55 \times 354.356 = 3384 \, \text{rev} / \text{min}$$

The rotational speed of the mold = N = 3384 rev/min.

Example 11.3 Determining the Centrifugal Force in Centrifugal Casting

By using the data in Examples 11.1 and 11.2, determine the centrifugal force acting on the molten metal during the casting process.

Solution

Mass of copper = m = 1 kg, G_f = 64, F =?

By using Eq. 11.2,

$$G_f = \frac{F}{m \cdot g}$$

Or

$$F = G_f * m * g = 64 \times 1 \times 9.81 = 627.84 \, \text{N}$$

The centrifugal force = F = 627.84 N.

Example 11.4 Determining the Tangential Velocity of Molten Metal in TCC

By using the data in Examples 11.1–11.3, determine the tangential speed of the molten metal during the centrifugal casting process.

Solution

N = 3384 rev/min,

R = Inside radius of the mold = Outside radius of the casting = $D/2$ = $\dfrac{0.01}{2}$ m = 0.005 m.

By using Eq. 11.5,

$$v = \frac{\pi \cdot R \cdot N}{30} = \frac{\pi \times 0.005 \times 3384}{30} = 1.77 \, \text{m} / \text{s}$$

The tangential speed of the molten metal = v = 1.77 m/s.

Example 11.5 Verifying the Tangential Speed of Molten Metal Using Definition of F

Verify the answer in Example 11.4 by using the basic numerical definition of centrifugal force.

Solution

The centrifugal force is numerically defined by the relationship given in Eq. 11.1 :

$$F = \frac{m \cdot v^2}{R}$$

Or

$$v^2 = \frac{F * R}{m} = \frac{627.84 \times 0.005}{1} = 3.135$$

$$v = \sqrt{3.135} = 1.77 \, \text{m/s}$$

The answer in Example 11.4 stands verified.

Example 11.6 Determining the G-Factor in Centrifugal Casting

It is required to produce a brass tube using a horizontal true centrifugal casting process. The length of the tube is 1.4 m with outside diameter of 14.5 cm and inside diameter of 12.0 cm. Determine the G-factor, if the rotational speed of the mold will be 1000 rev/min.

Solution

D = outside diameter of casting = 14.5 cm = 0.145 m, N = 1000 rev/min, g = 9.81 m/s^2, G_f =?

$$N = \frac{30}{\pi} * \sqrt{\frac{2 * G_f * g}{D}}$$

$$1000 = \frac{30}{\pi} * \sqrt{\frac{2 * G_f \times 9.81}{0.145}} = 9.55 \times \sqrt{135.3 \, G_f}$$

$$9.55 \times \sqrt{135.3 \, G_f} = 1000$$

$$\sqrt{135.3 \, G_f} = 104.712$$

$$135.3 \, G_f = 10{,}964.6$$

$$G_f = 81$$

Example 11.7 To Assess the Success of a Centrifugal Casting Operation
By using the data in Example 11.6, assess the likelihood of the success of the casting operation.

Solution
$G_f = 81$.

Since G_f should be in the range of 60–70, it is unlikely that the casting operation is successful.

Example 11.8 Determining the Rotational Speed of the Mold in Vertical TCC
It is required to cast a metal tube of length 20 cm by vertical TCC. The inside diameter at the top and bottom is 16 and 12 cm, respectively. Determine the rotational speed of the mold for the casting process.

Solution
$L = 20$ cm $= 0.20$ m, $R_t = \frac{1}{2} \times 16 = 8$ cm $= 0.08$ m, $R_b = \frac{1}{2} \times 12 = 6$ cm $= 0.06$ m, $N = ?$

By using Eq. 11.8,

$$N = \frac{30}{\pi} * \sqrt{\frac{2 * g * L}{\left(R_t^2 - R_b^2\right)}} = \frac{30}{\pi} \times \sqrt{\frac{2 \times 0.2}{\left(0.08^2 - 0.06^2\right)}} = 9.55 \times \sqrt{\frac{0.4}{\left(0.0064 - 0.0036\right)}}$$

$$N = 9.55 \times \sqrt{\frac{0.4}{0.0028}} = 114.14 \, \text{rpm}$$

The rotational speed of the mold $= 114.14$ rev/min.

Example 11.9 Determining the Inside Diameter at the Top of the Tube for Vertical TCC
A vertical TCC process is used to produce tubes that are 180 mm long and 160 mm in outside diameter. The rotational speed of the mold, during solidification, is 600 rev/min. Determine the inside diameter at the top of the tube, if the inside diameter at the bottom is 130 mm.

Solution
$L = 180$ mm $= 0.18$ m, $N = 600$ rev/min, $R_b = \frac{1}{2} \times 130 = 65$ mm $= 0.065$ m, $R_t = ?$

$$N = \frac{30}{\pi} * \sqrt{\frac{2 * g * L}{\left(R_t^2 - R_b^2\right)}}$$

$$600 = 9.55 \times \sqrt{\frac{2 \times 9.81 \times 0.18}{\left(R_t^2 - 0.065^2\right)}}$$

$$\sqrt{\frac{2\times9.81\times0.18}{\left(R_t^2-0.065^2\right)}}=62.827$$

$$\sqrt{\frac{3.532}{\left(R_t^2-0.065^2\right)}}=62.827$$

$$\frac{3.532}{\left(R_t^2-0.065^2\right)}=3947.23$$

$$\left(R_t^2-0.065^2\right)=0.0008948$$

$$R_t^2=0.0008948+0.0004225=0.0051198$$

$$R_t=\sqrt{0.0051198}=0.0715\,\text{m}=71.5\,\text{mm}$$

The inside diameter at the top of the tube $= 2\,R_t = 2 \times 71.5 = 143$ mm.

Example 11.10 Determining the Rotational Speed of the Driven Pulley

An open-belt drive system is used to rotate the mold of a centrifugal casting machine, as shown in Fig. 11.7. Determine the rotational speed of the driven pulley.

Solution

With reference to Figs. 11.6 and 11.7, $n_1 = 1000$ rpm, $d_1 = 100$ mm, $d_2 = 50$ mm, $n_2 = ?$

By using the modified form of Eq. 11.9,

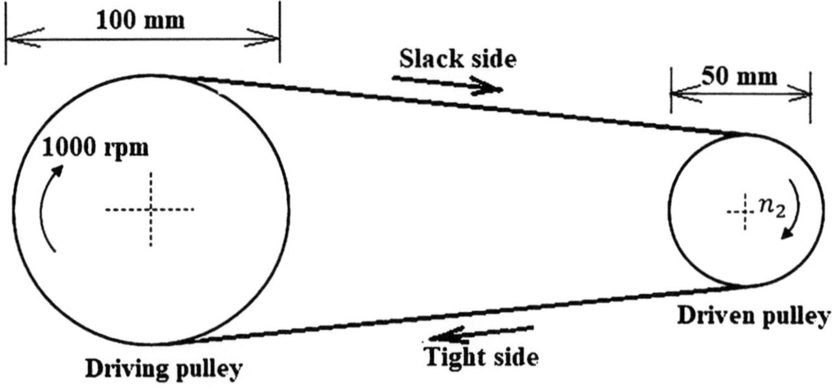

Fig. 11.7 Belt-pulley system for Example 11.7

$$n_2 = \frac{n_1 * d_1}{d_2} = \frac{1000 \times 100}{50} = 2000\,\text{rpm}$$

The rotational speed of the driven pulley = n_2 = 2000 rev/min.

Example 11.11 Determining the Peripheral Velocity of the Belt on the Driving Pulley
By using the data in Example 11.10, determine the peripheral velocity of the belt on the driving pulley.

Solution
n_1 = 1000 rpm, d_1 = 100 mm, v_1 =?
   By using Eq. 11.10,

$$v_1 = \frac{\pi \cdot d_1 \cdot n_1}{60} = \frac{\pi \times 100 \times 1000}{60} = 5236\,\text{m/s}$$

The peripheral velocity of the belt on the driving pulley = v_1 = 5236 m/s.

Example 11.12 Determining the Peripheral Velocity of the Belt on the Driven Pulley
By using the data in Example 11.10, determine the peripheral velocity of the belt on the driven pulley.

Solution
d_2 = 50 mm, n_2 = 2000 rpm, v_2 =?

Solution
By using Eq. 11.11,

$$v_2 = \frac{\pi \cdot d_2 \cdot n_2}{60} = \frac{\pi \times 50 \times 2000}{60} = 5236\,\text{rpm}$$

The peripheral velocity of the belt on the driven pulley = v_2 = 5236 rpm.

Example 11.13 Proving the Design Relationship for Open-Belt Drive System
By using the data in Examples 11.11 and 11.12, prove Eq. 11.9.

Solution
v_1 = 5236 rpm = v_2.

$$v_1 = v_2$$

By using Eqs. 11.10 and 11.11,

$$\frac{\pi \cdot d_1 \cdot n_1}{60} = \frac{\pi \cdot d_2 \cdot n_2}{60}$$

   Or

$$n_1 \cdot d_1 = n_2 \cdot d_2$$

Or

$$\frac{n_1}{n_2} = \frac{d_2}{d_1}$$

Equation 11.9 stands proved.

Example 11.14 Determining the Length of Belt for Operating Centrifugal Casting Machine
By using the data in Example 11.10, determine the total length of the belt. The distance between the centers of the two pulleys is 1.5 m.

Solution
$d_1 = 100$ mm, $d_2 = 50$ mm, $c = 1.5$ m = 1500 mm, $L = ?$
By using Eq. 11.12,

$$L = 2 \cdot c + \frac{\pi}{2} \cdot \left(d_1 + d_2\right) + \frac{\left(d_1 - d_2\right)^2}{4 \cdot c} = \left(2 \times 1500\right) + \frac{\pi}{2} \times \left(100 + 50\right) + \frac{\left(100 - 50\right)^2}{4 \times 1500}$$

$$L = 3000 + 235.62 + 0.417 = 3236 \, \text{mm} = 3.236 \, \text{m}$$

The total length of the belt = 3.236 m.

Questions and Problems
11.1. Encircle the most appropriate answers for each of the following statements.

(1) What is the range of the rotational speed of the mold in centrifugal casting?
   (a) 300–3000 rpm, (b) 100–300 rpm, (c) 3000–5000 rpm, (d) 5000–10,000 rpm
(2) Which component is not generally cast by centrifugal casting?
   (a) Pipe, (b) bushing, (c) tubes, (d) blades
(3) What should be the wall-thickness range of hollow cylinders in true centrifugal casting?
   (a) 1.0–2.5 mm, (b) 2.5–125 mm, (c) 125–400 mm, (d) 400–500 mm
(4) What is the range of G-factor in true centrifugal casting?
   (a) 40–50, (b) 50–60, (c) 60–70, (d) 70–80
(5) Which process is the most suitable for casting small parts (e.g. jewelry, bushes, etc.)?
   (a) True centrifugal casting (TCC), (b) centrifuge casting, (c) semi-centrifugal casting
(6) In which process, the inside profile of the casting wall takes the shape of a parabola?

(a) Vertical TCC, (b) horizontal TCC, (c) semi-centrifugal casting, (d) centrifuging

(7) Which process is the most suitable for casting a spoked wheel?

(a) Vertical TCC, (b) horizontal TCC, (c) semi-centrifugal casting, (d) centrifuging

(8) Which type of centrifugal casting is the most commonly practiced process?

(a) Vertical TCC, (b) horizontal TCC, (c) semi-centrifugal casting, (d) centrifuging

(9) Which force is responsible for the accumulation of impurities near the center of the wheel in semi-centrifugal casting?

(a) Centrifugal force, (b) metallostatic force, (c) buoyancy force, (d) centripetal casting

(10) Which type of casting involves the inside radii at the top and bottom of the casting?

(a) Horizontal TCC, (b) vertical TCC, (c) semi-centrifugal casting, (d) centrifuging

11.2. List the advantages and limitations of centrifugal casting.

11.3. Explain the true centrifugal casting (TCC) process with the aid of a sketch.

11.4. Why does the outer skin of centrifugally cast component have a high strength?

11.5. Differentiate between the horizontal TCC and the vertical TCC processes.

11.6. Describe the semi-centrifugal casting process with the aid of a sketch.

11.7. Why is the center of a semi-centrifugally cast component removed?

P-11.8. A 100-cm-long steel tube with outside diameter of 16 mm and inside diameter of 10 mm is to be produced by true centrifugal casting. Calculate the mass of steel required to be poured for the casting process. Ignore all allowances. The density of steel is 7.85 g/cm^3.

P-11.9. By using the data in P-11.8, determine the rotational speed of the mold for the TCC, if G-factor is 66.

P-11.10. By using the data in P-11.8–P-11.9, determine the centrifugal force acting on the molten metal during the casting process.

P-11.11. By using the data in P-11.8–P-11.10, determine the tangential speed of the molten metal during the centrifugal casting process.

P-11.12. It is required to cast a metal tube of length 25 cm by vertical TCC. The inside diameter at the top and bottom is 18 and 15 cm, respectively. Determine the rotational speed of the mold for the casting process.

P-11.13. A vertical TCC process is used to produce tubes that are 160 mm long and 120 mm in outside diameter. The rotational speed of the mold, during solidification, is 800 rev/min. Determine the inside diameter at the top of the tube, if the inside diameter at the bottom is 100 mm.

P-11.14. In a two-pulley-belt (open-belt drive) system, the rotational speed of the driving pulley is 800 rev/min, the diameter of the driving pulley is 90 mm, and the diameter of the driven pulley is 60 mm. Determine the rotational speed of the driven pulley.

P-11.15. By using the data in P-11.14, determine the peripheral velocity of the belt on the (a) driving pulley and (b) driven pulley.

P-11.16. By using the data in P-11.14, determine the total length of the belt. The distance between the centers of the two pulleys is 1.2 m.

References

Campbell, J. (2011). *Complete casting handbook: Metal casting processes, techniques and design.* Butterworth-Heinemann/Elsevier Science Publications.

Khurmi, R. S., & Gupta, J. K. (2005). *A textbook of machine design.* Eurasia Publishing House.

Oyewole, A., & Sunday, A. M. (2011). Design and fabrication of a centrifugal casting machine. *International Journal of Engineering Science and Technology, 3*(11), 8204–8210.

Romanoff, P. (1981). *The complete handbook of centrifugal casting.* Tab Books.

Chapter 12
Testing, Casting Defects, and Quality Assurance

12.1 Casting Processing Conditions and Mechanical Testing

Mechanical testing of casting is of great significance for quality assurance (QA) in foundry practice. The mechanical properties and behavior of a casting strongly depend on the processing conditions, such as the pouring rate, cooling rate, etc. For example, rapid solidification/cooling results in a fine-grained microstructure associated with higher strength, whereas solidification with a slow cooling rate produces a coarse-grained microstructure indicating a lower-strength material. It is, therefore, important in industrial practice to conduct mechanical testing of the castings to ensure that the products is defect-free and meet mechanical properties' specification, such as tensile strength, hardness, impact toughness, and the like.

Mechanical testing may be either destructive testing or non-destructive testing (NDT). The destructive mechanical testing involves cutting out a piece from the casting to prepare a test specimen/sample. The examples of destructive testing include tensile testing, hardness testing, impact testing, fatigue testing, creep testing, etc. On the other hand, NDT involves no destruction of the casting. It includes visual examination, magnetic particle test, dye penetrant test, ultrasonic testing, radiographic examination, and the like.

12.2 Destructive Mechanical Testing

12.2.1 Tensile Testing, σ-ε Curves, and Mechanical Properties

Prior to tensile testing, it is necessary to produce a tensile test specimen, as shown in Fig. 12.1. The dimensions of the test specimen must be in accordance with the standards prescribed by the American Society of Testing of Materials (ASTM,

© The Author(s), under exclusive license to Springer Nature
Switzerland AG 2025

Z. Huda, *Metal Casting Engineering*, Mechanical Engineering Series,
https://doi.org/10.1007/978-3-031-84620-5_12

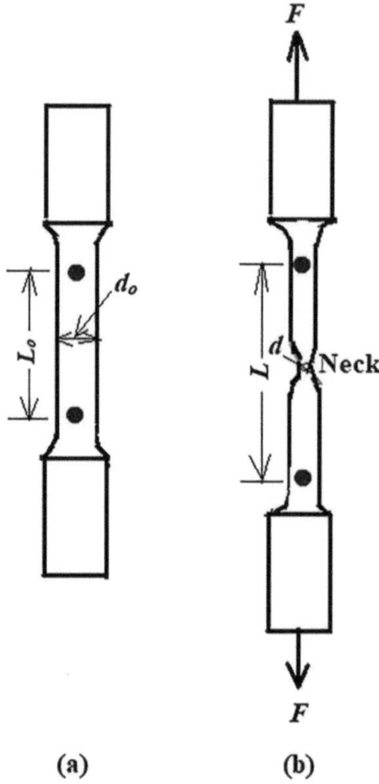

Fig. 12.1 Tensile test specimen before loading (**a**), and after loading (**b**)

2013). The original gauge length (L_o) and the original diameter (d_o) are measured and recorded (see Fig. 12.1a).

The tensile test is conducted by use of a tensile testing machine or universal testing machine; the latter can be used for both tension and compression tests. In tensile testing, the specimen is firmly held in place by the grips of the testing machine (see Fig. 12.2). One end of the specimen is held firm, while the other end is pulled by applying a tensile force within the specimen. An electronic device (extensometer) is mounted on the specimen for measuring the specimen extension. The extensometer must be removed once the specimen approaches its proportional limit else it will be damaged when the specimen breaks. Once the material has fractured, the specimen is re-assembled, and the gauge length at failure (L_f), and the final cross-sectional area (A_f) are measured (see Fig. 12.1b).

By using the force and dimensional changes data, the engineering stress (σ_eng) and engineering strain (ε_eng) can be calculated by using the following mathematical relationships:

$$\sigma_\mathrm{eng} = \frac{F}{A_\mathrm{o}} \qquad\qquad (12.1)$$

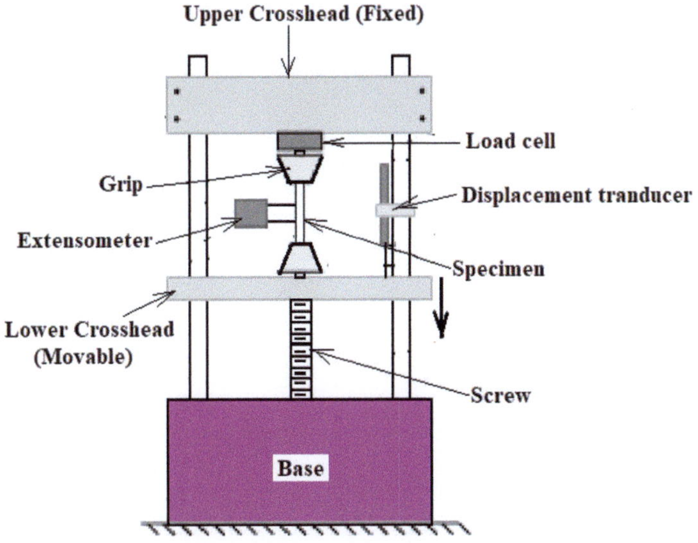

Fig. 12.2 Tensile testing – *working principle*

$$\varepsilon_{eng} = \frac{\Delta L}{L_o} = \frac{L_f - L_o}{L_o} \tag{12.2}$$

where σ_{eng} is the engineering stress, MPa; F is the applied force, N; A_o is the original cross-sectional area of the specimen, mm^2; ε_{eng} is the engineering strain; L_o is the original length of the specimen (or distance between gage marks), mm; and L_f is the final gage length just before fracture, mm (see Examples 12.1–12.3). The data, obtained from the tensile test, is used to calculate tensile mechanical properties (see Eqs. 12.3–12.10).

Stress-Strain Curve The tensile testing results in deformation (elongation) of the specimen until it breaks. A graphical representation of the relationship between load applied and deformation of the material is expressed as a stress-strain curve (Fig. 12.3).

The first stage of the stress-strain curve (Fig. 12.3) indicates elastic behavior of material; here the stress is proportional to the elastic strain up to the proportional limit (point A), i.e., *Hook's Law* is obeyed from O to A. This linear elastic behavior is expressed as:

$$\sigma = E\varepsilon_e \tag{12.3}$$

where σ is the *stress* in the elastic range, ε_e is the *elastic strain*, and the constant of proportionality E is called *Young's modulus* or modulus of elasticity.

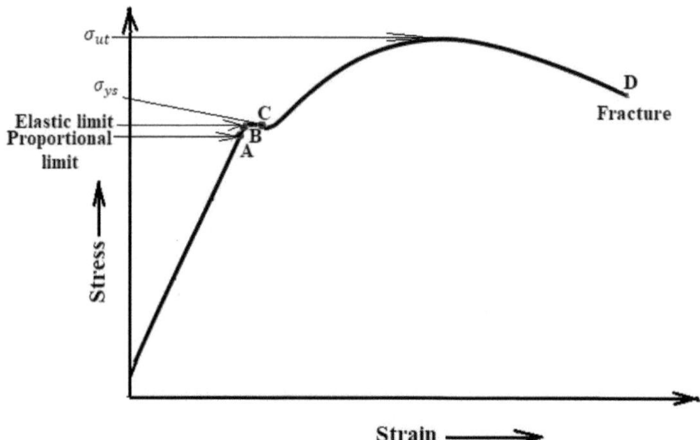

Fig. 12.3 Stress-strain curve for a ductile metal (σ_{ys} = yield stress, σ_{ut} = ultimate tensile strength)

Table 12.1 Tensile mechanical properties of some materials

Material	Yield strength (σ_{ys}), MPa	Tensile strength (σ_{ut}), MPa	% Elongation (in L_0 = 50 mm)	Young's modulus (E)
Aluminum	35	90	40	70 GPa
Copper	70	200	45	110 GPa
Steels	220–1000	350–2000	12–30	M.S 207 GPa
Cast irons	100–150	300–1000	0–7	162–170 GPa
Nickel	140	480	40	207 GPa
Magnesium alloys	170–200	300–320	10–13	45 GPa
Titanium	450	520	25	120 GPa

Beyond the proportional limit, the straight line becomes a curve. The stress at the elastic point *B* is called the *elastic limit* (see Fig. 12.3). Then, the material starts to yield at the yield point *C*. The stress-strain relationship for yielding can be expressed as:

$$\sigma_{ys} = E\epsilon_{yield} \tag{12.4}$$

where σ_{ys} is the yield strength (MPa) and ϵ_{yield} is the elastic strain taken to yielding.

The modulus of elasticity (E) values of some metals and alloys are listed in Table 12.1. Young's modulus may be computed by rewriting Eq. 12.3 as follows:

$$E = \frac{\text{stress}}{\text{Elastic strain}} = \frac{\dfrac{F}{A_o}}{\dfrac{\delta l}{L_o}} = \frac{FL_o}{A_o(\delta l)} \tag{12.5}$$

where F is the force up to proportional limit (N), δl is the elastic elongation (mm), and E is the Young's modulus (MPa) (see Examples 12.4 and 12.5).

Beyond the yield point (C), the material *strain hardens*, i.e., there is an increase in stress with increasing strain (see Fig. 12.3). The stress-strain curve enables us to determine both strength and ductility properties. For example, the strength properties that can be determined by the stress-strain curve include the yield strength (σ_{ys}), the ultimate tensile strength (σ_{UTS}), and the fracture strength or breaking strength (σ_b). The stress-strain curve, as obtained from the tensile test of a mild steel (AISI-1020) sample, is shown in Fig. 12.4, which enables us to determine the tensile mechanical properties of AISI-1020 steel (see Examples 12.6–12.10).

The tensile test data enables us to determine strength properties by using the following formulas:

$$\sigma_{ys} = \frac{F_y}{A_o} \tag{12.6}$$

$$\sigma_{ut} = \frac{F_{max}}{A_o} \tag{12.7}$$

$$\sigma_b = \frac{F_b}{A_o} \tag{12.8}$$

where F_y is the force at the yield point, F_{max} is the maximum force during the tensile test, and F_b is the force at break (see Examples 12.11–12.13).

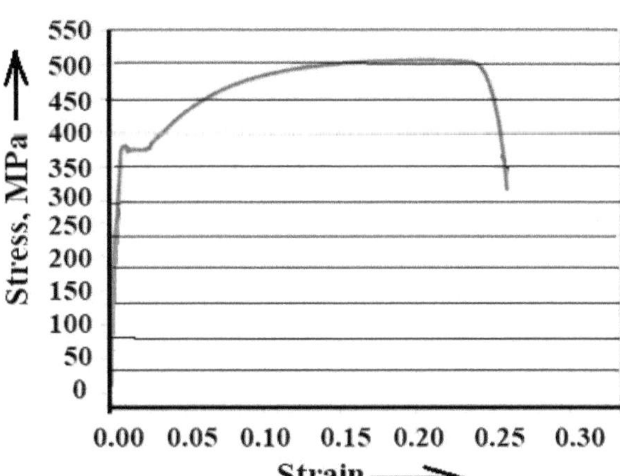

Fig. 12.4 The tensile stress-strain curve for AISI-1020 steel ($l_o = 50$ mm

The stress-strain curve also enables us to determine ductility properties. The percent elongation and the percent reduction in area (% RA) can be determined by:

$$\%\text{Elongation} = \frac{\Delta L}{L_\text{o}} \times 100 = \frac{L_\text{f} - L_\text{o}}{L_\text{o}} \times 100 \tag{12.9}$$

$$\%\text{Reduction in Area} = \%\text{RA} = \frac{A_\text{o} - A_\text{f}}{A_\text{o}} \times 100 \tag{12.10}$$

The significance of Eqs. 12.9 and 12.10 is illustrated in Examples 12.14–12.18. In order to determine the compressive mechanical properties, a universal testing machine is used (see Examples 12.19–12.21). The tensile mechanical properties of some engineering materials are presented in Table 12.1.

12.2.2   Hardness Testing

12.2.2.1   Hardness Testing Principle and Types

Hardness refers to the ability of a material to resist plastic deformation by indentation (penetration), scratching, or abrasion. Hardness is generally measured as the indentation hardness, which is defined as the resistance of a metal to penetration or indentation. *Hardness testing* involves the use of a pointed or rounded indenter that is pressed into a surface under a substantially static load (Pelleg, 2013). There are four commonly used methods of hardness testing: (a) Brinell hardness test, (b) Rockwell hardness test, (c) Vickers test, and (d) Knoop hardness test. These hardness testing methods are briefly explained in the following subsections.

12.2.2.2   Brinell Hardness Test

The Brinell hardness test involves indentation of the test material by use of a 10-mm-diameter hardened steel/carbide ball that is subjected to a specified load, F (see Fig. 12.5a). The full load of 3000 kgf is applied for 10–15 s on hard materials, whereas a lower force, in the range of 500–1500 kgf, is applied for about 30 s on soft materials. A low-power microscope is used to measure the diameter of the indentation produced in the test material (see Fig. 12.5b).

The Brinell harness number (BHN) can be calculated by dividing the load applied (F) by the surface area of the indentation by using the following formula:

$$\text{BHN} = \frac{F}{\frac{\pi}{2}\left(D - \sqrt{D^2 - D_\text{i}^2}\right)} \tag{12.11}$$

where D and D_i are the diameters of indenter and indentation, respectively (see Example 12.22).

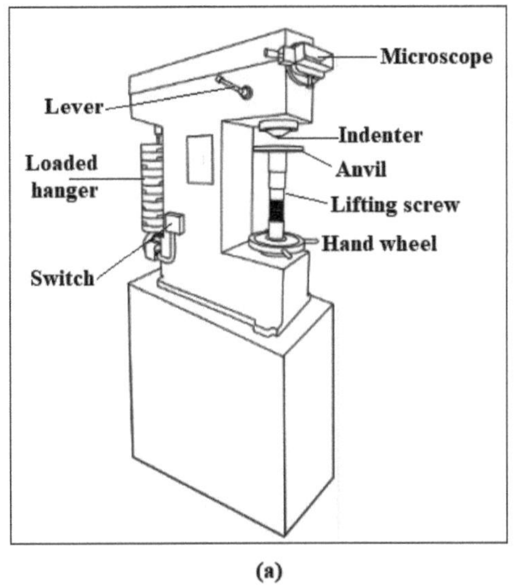

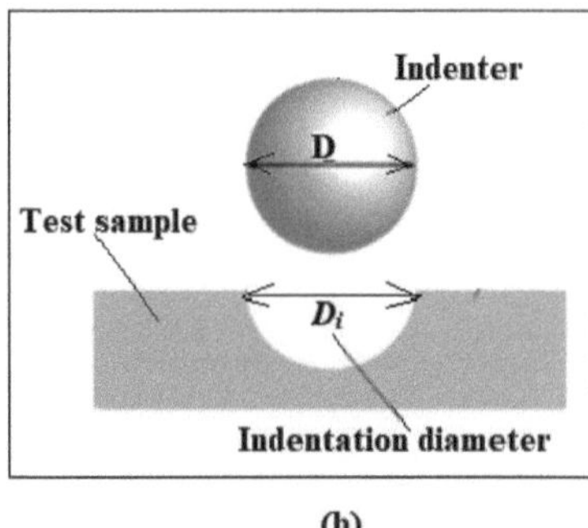

Fig. 12.5 Schematic of Brinell hardness testing equipment (**a**), and principle (**b**)

Table 12.2 Brinell hardness numbers of some alloys at 25 °C

Material	Brasses and Al alloys	Carbon steels	Cast irons	Hardened steels
BHN	60–100	130–235	350–415	800–900

A well-defined BHN indicates the test conditions. For example, 120 HB 10/1000/25 means that Brinell hardness number of 120 was obtained using a 10-mm-diameter hardened steel ball with a 1000 kgf load applied for 25 s (see Example 12.23). For testing of extremely hard metals, the steel ball is replaced by a tungsten carbide (WC) ball. Table 12.2 lists BHN values of some commonly used alloys.

For most steels, there exists a relationship between Brinell hardness and strength according to:

$$\sigma_{ut} = 4.35 * \left(\text{BHN}\right) \tag{12.12}$$

where σ_{ut} is the ultimate tensile strength, MPa (see Example 12.24).

12.2.2.3 Rockwell Hardness Test

The Rockwell hardness testing involves indenting the test material using a hardened steel ball indenter or a diamond cone-shaped Brale indenter (see Fig. 12.6a). Firstly, the indenter is forced into the test material under a minor load *F0,* usually 10 kgf (see Fig. 12.6b). On reaching equilibrium, an indicating device is set to a datum position. While the minor load is kept on, an additional major load (*F1*) is applied

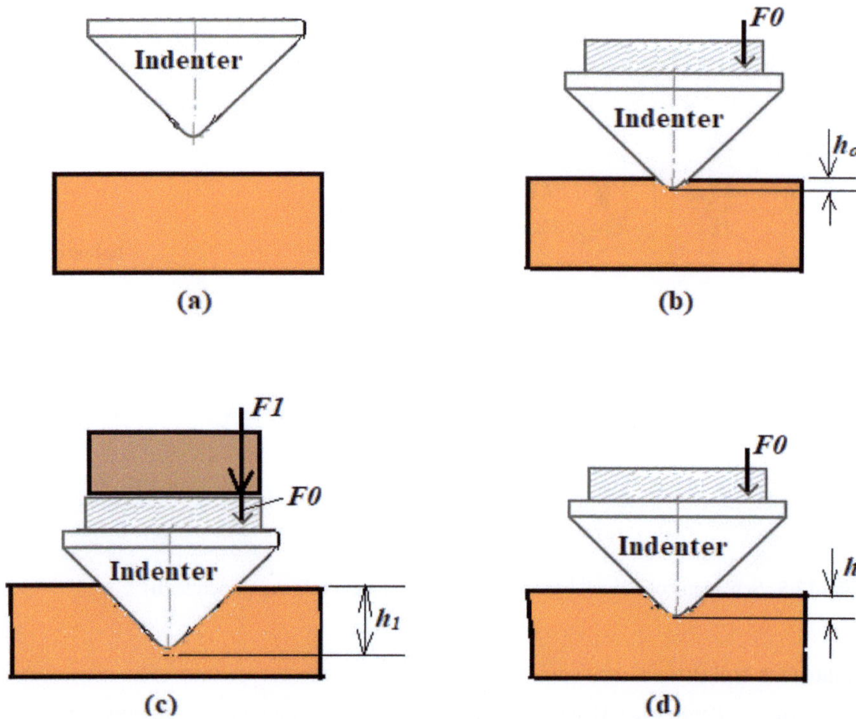

Fig. 12.6 Rockwell hardness testing principle; (**a**) an indenter is positioned above the test sample, (**b**) force F_0 is applied to sample resulting in indentation of depth h_0, (**c**) additional force F_1 is applied increasing the inentation to depth h_1, (**d**) the force F_1 is removed to obtain the final indentation of depth h

that results in an increase in the penetration (Fig. 12.6c). Once equilibrium has again been reached, the major load is removed, but the minor load is still kept maintained. The removal of the major load results in a reduction of the depth of penetration h (see Fig. 12.6d).

The Rockwell hardness number (HR) can be calculated by the following formula:

$$\text{HR} = R - \frac{h}{s} \tag{12.13}$$

where h is the permanent increase in the depth of penetration, s is usually 0.002 mm, and R is a constant depending on the form of indenter; $R = 100$ units for diamond indenter, and $R = 130$ units for steel ball indenter (see Example 12.25).

The Rockwell hardness testing indicates the final test results directly on a dial which is calibrated with a series of scales. This testing method is employed for rapid testing of finished materials and products in foundries and other industries. In general, scales A–C are used for metals and alloys (see Table 12.3). Rockwell hardness numbers for steels range from 20 HRC (for mild steel) to 80 HRC (for nitrided steels). Brasses and aluminum alloys have hardness numbers in the range of 20–60 HRB.

Table 12.3 Rockwell hardness scales and loads for typical applications

Scale	Indenter	Minor load $F0$, kgf	Major load $F1$, kgf	R	Typical applications
A	Diamond cone	10	50	100	Cemented carbides
B	$\frac{1}{16}$″ steel ball	10	90	130	Copper alloys, mild steel, aluminum alloys, malleable irons
C	Diamond cone	10	140	100	Steels, hard cast irons, case hardened steel, and other materials harder than 100 HRB

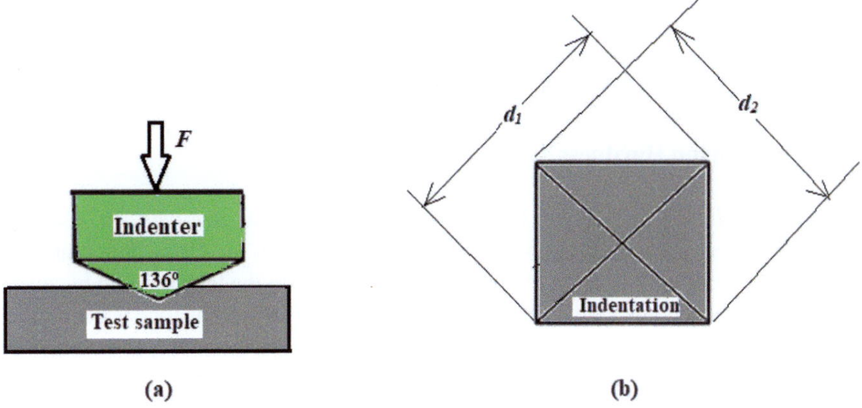

Fig. 12.7 Vickers hardness testing principle (**a**), and the indentation diagonals (**b**)

12.2.2.4 Vickers Hardness Test

The Vickers hardness testing involves indenting the test material using a diamond indenter. The indenter is in the form of a right pyramid with a square base. The angle between the opposite faces is 136° (see Fig. 12.7a). The indenter is subjected to a load in the range of 1–100 kgf. The full load is generally applied for 10–15 s. The two diagonals of the indentation on the surface of the material after removal of load are measured using a microscope (see Fig. 12.7b).

The indentation diagonals d_1 and d_2 (in mm) enable us to calculate d, which is the arithmetic mean of the two diagonals (see Fig. 12.7b). Vickers hardness number (VHN) is calculated by:

$$\text{VHN} = \frac{2F \sin \dfrac{136}{2}}{d^2} = \frac{1.854F}{d^2} \tag{12.14}$$

where F is the force applied to the indenter (kgf) (see Examples 12.26 and 12.27).

The Vickers hardness number should be reported in the form: 700 HV/60, which means a Vickers hardness of 700 was obtained using a force of 60 kgf. The Vickers hardness test has several advantages, including high accuracy in readings, and the

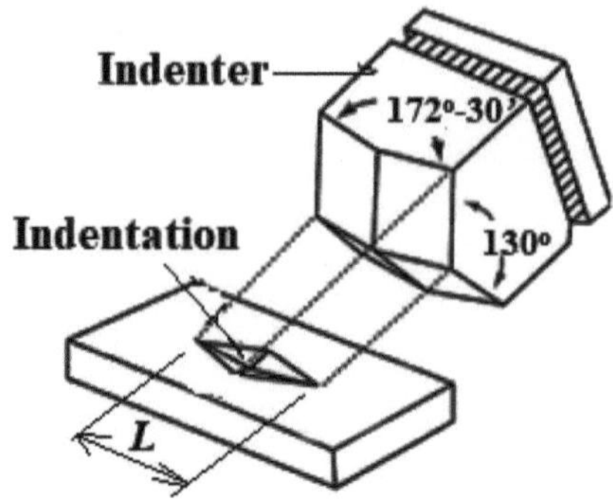

Fig. 12.8 Knoop hardness testing principle

use of just one type of indenter for all types of metals/alloys and surface treated materials.

12.2.2.5   Knoop Hardness Test

The *Knoop hardness test*, or Knoop microhardness test, involves the use of a rhombic-based pyramidal diamond indenter that forms an elongated diamond-shaped indentation. The diamond indenter is pressed into the test material by applying an accurately controlled force, P, in the range of 1–1000 g. The load is maintained for a specific dwell time in the range of 10–15 s. Then, the indenter is removed, leaving an elongated diamond-shaped indentation in the test material. An optical microscope is used to measure the longest diagonal (L) of the diamond shaped indentation (see Fig. 12.8).

The Knoop hardness number (KHN) is calculated by using the following formula:

$$\text{KHN} = \frac{14.23\,P}{L^2} \tag{12.15}$$

where P = force (kgf), and L = the longest diagonal of indentation (mm) (see Example 12.28).

The Knoop hardness number normally ranges from HK60 to HK600 for brasses, aluminum alloys, and soft steels. Cutting tools and nitride steels have hardness from HK700 to HK1000.

12.2.3   *Impact Testing*

The *impact testing* involves the use of a pendulum with a known mass at the end of its arm. The pendulum swings down and strikes a notched specimen, thereby fracturing it (see Fig. 12.9a). There are two principal methods for impact testing: (a) Charpy impact test, and (b) Izod impact test. In the *Charpy impact test*, the test

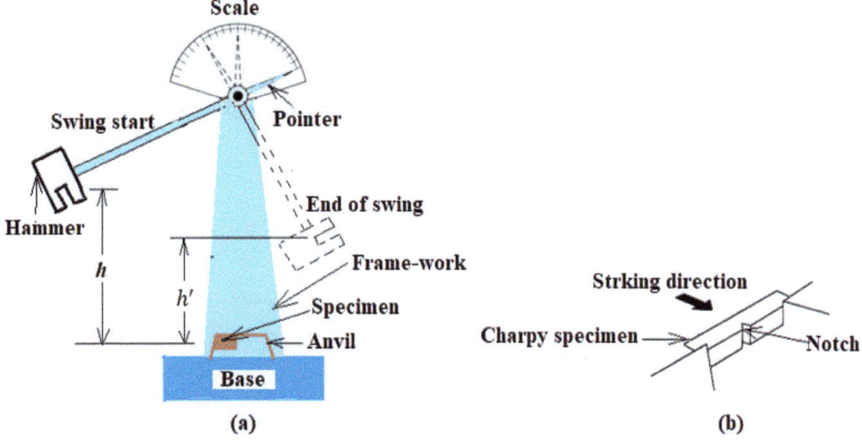

Fig. 12.9 Impact testing principle (**a**), and Charpy test specimen (**b**)

Table 12.4 Impact energies of some alloys at 25 °C

Material	6061 *Al* alloy	1045 cold-worked steel	1045 normalized steel	1095 normalized steel	304 stainless steel
Charpy impact energy (J)	35	15	74	50	130

specimen is securely held in a horizontal position, and the notch is positioned facing away from the striker (see Fig. 12.9b). In the *Izod impact test*, the notched specimen is held in a vertical position with the notch positioned facing the striker.

The impact energy can be determined by:

$$U_{\mathrm{i}} = mg\left(h - h'\right) \tag{12.16}$$

where U_{i} is the impact energy of the specimen's material (J), m is the mass of the pendulum (kg), g is the gravitational acceleration ($g = 9.81$ m/s^2), and $(h - h')$ is the difference between the initial and final heights of the pendulum (see Example 12.29).

The impact energies of some engineering materials are presented in Table 12.4.

12.2.4 Fatigue Testing: Cyclic Loading and Stress Cycle Parameters

Fatigue refers to the weakening or failure of a material due to the formation and propagation of cracks resulting from a repetitive or cyclic load. The examples of components that usually fail because of fatigue include: rotating shafts, aircraft wings, springs, turbine blades, gears, etc.

Stress Cycle Parameters In components subjected to cyclic loading, it is important to determine the various stress cycle parameters, which include: (a) mean stress, (b) stress range, (c) stress amplitude, (d) stress ratio, and (e) amplitude ratio. The *mean stress* (σ_m) is the average of maximum and minimum stresses in a cycle. Numerically (Huda, 2020):

$$\sigma_m = \frac{\sigma_{max} + \sigma_{min}}{2} \tag{12.17}$$

where σ_{max} is the maximum stress, and σ_{min} is the minimum stress. In case of alternating stress cycle, the mean stress is zero, 0 (see Example 12.30).

The stress range (σ_r) is the difference of the maximum stress and the minimum stress.

$$\sigma_r = \sigma_{max} - \sigma_{min} \tag{12.18}$$

The stress amplitude (σ_a) in the alternating stress cycle is one-half of the stress range.

$$\sigma_a = \frac{\sigma_r}{2} \tag{12.19}$$

The stress ratio (R_s) is the ratio of the minimum stress to the maximum stress. Mathematically,

$$R_s = \frac{\sigma_{min}}{\sigma_{max}} \tag{12.20}$$

The *amplitude ratio* (R_a) is the ratio of the stress amplitude to the mean stress. Mathematically,

$$R_a = \frac{\sigma_a}{\sigma_m} \tag{12.21}$$

where σ_a is the stress amplitude, and σ_m is the mean stress. The significance of Eqs. 12.18–12.21 is illustrated in Examples 12.31–12.34.

Fatigue Testing Fatigue testing is the mechanical testing that involves the application of cyclic loading to a test piece (Huda et al., 2020). In fatigue testing, a cylindrical test piece is subjected to compressive and tensile stresses as the test-piece is simultaneously bent and rotated by use of a motor and bearings (see Fig. 12.10). The results, from the fatigue test, are plotted in the form of an S-N curve (stress amplitude vs. number of cycles to failure curve) (Huda, 2022).

It is important to define three important terms: *fatigue strength*, *fatigue limit*, and *fatigue life*. *Fatigue strength* is the stress amplitude at which failure occurs for a given number of cycles. *Fatigue life* (N_f) is the number of cycles to failure at a certain stress. *Fatigue limit or endurance limit* (S_e) is the stress level below which a test

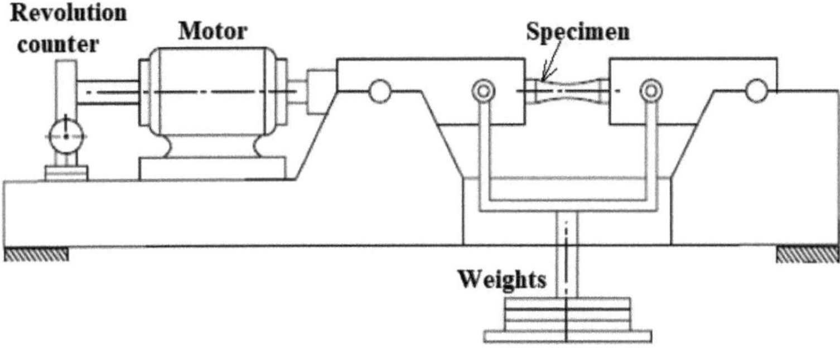

Fig. 12.10 Principle of rotating-bending fatigue testing

material can withstand cyclic stresses indefinitely without exhibiting fatigue failure. Most carbon steels and low-alloy steels exhibit a definite fatigue limit (S_e), which is generally about one-half of the tensile strength (σ_{ut}), i.e.:

$$S_e = \frac{\sigma_{ut}}{2} \tag{12.22}$$

Goodman's Law Goodman's law expresses a relationship between the stress amplitude, mean stresses, tensile strength, and fatigue strength, as follows:

$$\frac{\sigma_a}{S_e} + \frac{\sigma_m}{\sigma_{ut}} = 1 \tag{12.23}$$

where σ_a is the stress amplitude, S_e is the fatigue limit, σ_m is the mean stress, and σ_{ut} is the ultimate tensile strength (see Examples 12.35 and 12.36).

12.3   Non-destructive Testing (NDT) of Castings

12.3.1   What Is NDT of Castings? And Why Is It Important?

Non-destructive testing (NDT) is a group of testing processes allowing us to measure and ensure the conformity of a part without having to destroy it. The NDT techniques used on castings include: (a) visual examination, (b) magnetic particle testing, (c) dye penetrant test, (d) ultrasonic inspection, (e) Eddy current test, (f) radiographic inspection, and the like. The reason for conducting NDT of castings is explained in the following paragraph.

Casting defects may be either visible on the surface or contained within the body of the cast part. In industrial practice, customers often want assurance that their castings are free from internal defects. In such cases, NDT of castings is necessary.

Cooling leads to shrinkage that can cause cracking. The cracks may break the surface of the part or could be purely internal. Surface-breaking cracks are often extremely difficult to see with the unaided eye because they are very thin and the textured surface of the casting, especially with sand castings, makes them virtually invisible. In addition to internal cracks, other casting defects that require NDT to be conducted include porosity and inclusions.

12.3.2 Visual Examination and Magnetic Particle Testing

The visual examination involves the inspection by an unaided eye. For example, surface cracks can easily be detected by visual examination. However, some surface cracks may be too thin to be detected by an unaided eye, or they may be purely internal. In such a case, magnetic particle testing of ferrous casting is useful.

The *magnetic particle inspection* detects surface or near-surface issues on the casting. This process involves inducing a magnetic field via an electric current around the casting. When a discontinuity is present, the magnetic field is disrupted.

12.3.3 Dye Penetrant Testing

The *dye penetrant test* is an NDT method to detect surface-breaking defects in the test material. This test detects visible cracks, flaws, pores, leaks, and other defects on the material's surface.

Firstly, a liquid penetrant (e.g., a colored dye) is applied to the surface of the test object, covering the area of interest. As the penetrant has low surface tension, it seeps into any surface openings or defects by capillarity. Then, the penetrant is allowed to dwell on the surface for an agreed-upon time period. During this time, the penetrant infiltrates any defects, filling them. After the dwell period has passed, any excess penetrant is carefully removed using solvents or emulsifiers.

Finally, a developer (a white powder or a suspension) is applied to the surface. The developer helps to draw out the penetrant trapped within the defects, making them visible

12.3.4 Ultrasonic Inspection

12.3.4.1 Working Principle of Ultrasonic Inspection

The *ultrasonic testing* (UT) involves the use of an ultrasonic flaw detector, which emits short-pulse waves from a transducer to transmit through a test material (see Fig. 12.11). It is important to note that there is no backside reflection (in the scan display) when the defect is large. Most UT applications use short-pulse waves with

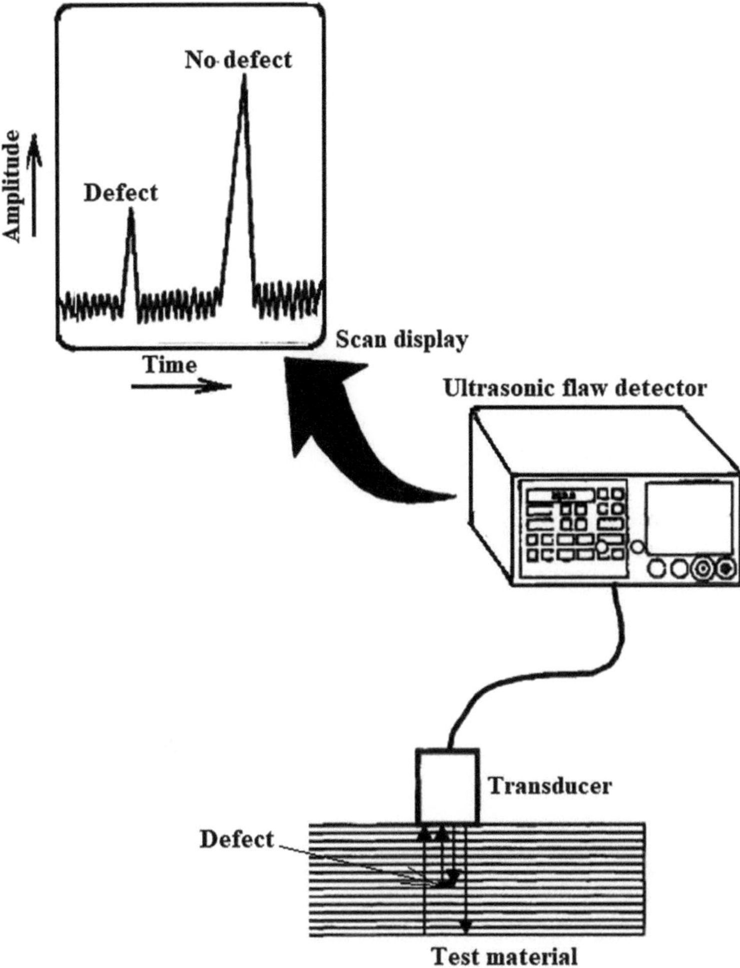

Fig. 12.11 Working principle of ultrasonic testing

frequencies usually in the range of 0.1–15 MHz. Ultrasonic flaw detectors are generally used to measure the wall dimensions of hollow castings. They can identify hidden discontinuities (e.g., porosity, inclusions, voids, and cracks) in castings.

12.3.4.2 Mathematical Modeling of Ultrasonic Inspection

The relationship between the incident and refracted angles of a wave, during ultrasonic testing, can be expressed by Snell's law, according to the following formula:

$$\frac{\sin\theta_1}{\sin\theta_2} = \frac{v_1}{v_2} \tag{12.24}$$

where θ_1 is angle of the incident wave, θ_2 is the angle of the refracted wave, v_1 is the velocity of incident wave, and v_2 is the velocity of refracted wave (see Example 12.37).

Near the ultrasonic transducer, there are significant fluctuations in the sound intensity due to the constructive and destructive interference of the waves which are issued from the transducer face. Because of acoustic variations, this field is called the *near field*. It is better to avoid positioning of the test-material in the *near field* because it can be extremely difficult to accurately evaluate flaws in the material when positioned within this area. The near field (N) can be determined by (CNE-ISU, 2024):

$$N = \frac{D^2 * f}{4v} \tag{12.25}$$

where is N is the near field (in), D is the transducer diameter (in), f is the wave frequency (Hz), and v is the sound velocity (in/s) (see Example 12.38).

Although a round transducer produces a cylindrical sound field in front of the transducer, the energy in the beam does not remain in a cylinder, but instead it spreads out as it propagates through the material. This phenomenon is referred to as *beam spread*. The *beam spread half angle* ($\theta°$) can be determined by (CNE-ISU, 2024):

$$\sin \theta = 1.2 * \left(\frac{v}{D * f} \right) \tag{12.26}$$

The significance of Eq. 12.26 is illustrated in Example 12.39.

12.3.5   Radiographic Examination

The radiographic NDT examination is generally conducted using either X-rays or gamma rays. In the radiographic testing, the part to be inspected is placed between the radiation source and a piece of radiation sensitive film. The part will stop some of the radiation where thicker. More dense areas will stop more of the radiation. The radiation that passes through the part will expose the film and forms a shadowgraph of the part. The film darkness (density) will vary with the amount of radiation reaching the film through the test material; darker areas indicate more exposure (higher radiation intensity) and lighter areas indicate less exposure (lower radiation intensity). This variation in the image darkness can be used to detect the presence of any flaw, defect, or discontinuity inside the material.

The relation between the radiation intensity and its distance can be expressed according to the inverse square law, as follows:

$$I_1 * d_1^2 = I_2 * d_2^2 \tag{12.27}$$

where I_1 and I_2 are the intensities at the distances d_1 and d_2 form the radiation source, respectively (see Example 12.40).

12.4 Casting Defects and Quality Assurance (QA)

12.4.1 Additional Processing in Metal Casting for QA

The control of quality is of the utmost importance in castings because customers want assurance that the castings, to be delivered to them, are free from defects. This is why the removal of a casting from the mold is usually not the final step in a metal casting process. There are additional processes required to obtain a good quality product. The additional steps in metal casting, for quality assurance (*QA*) include: trimming (removal of sprue, runner, riser, etc.), removing the core, surface cleaning, testing/inspection (both destructive testing and NDT), repair (if necessary), and heat treatment. The destructive testing results generally necessitate heat treatment for achieving the required mechanical properties (see Sect. 12.2). The NDT examinations would reveal surface and internal casting defects (see Sect. 12.3). Once a defect has been detected in a casting, the *QA* department must communicate their corrective recommendations to the production personnel for implementation.

The following defects are commonly found in sand castings: (a) misrun and cold shut, (b) hot tears, (c) shrinkage cavity, (d) blow holes, (e) pin holes, (f) metal penetration, and (g) mismatch. Additionally, porosity in aluminum die castings is also a common issue in industrial practice (Goodrich, 2008). These casting defects and their remedies are discussed in the following subsections. Although, most remedies relate to casting process itself, the production of a defect-free good-quality casting should be viewed from quality engineering point-of-view. For example, the inspection of machinery before its use (to ensure that it is fault-free) would help minimize casting defects.

12.4.2 Casting Defects and Their Remedies

12.4.2.1 Misrun and Cold Shut Defects

The misrun casting defect refers to an incomplete filling of mold cavity (see Fig. 12.12). The *cold shut* defect is the premature solidification of the melt during casting. The reasons of *misrun* and cold-shut defects may be one or more of the following: (a) insufficient metal supply to mold cavity, (b) too–low melt temperature, (c) improperly designed gates, or (d) too large length to thickness ratio (L/t) of the casting.

The misrun and cold shut defects can be prevented by one or more of the following casting techniques: (a) ensuring sufficient metal supply to the mold cavity, (b) increasing the melt temperature (superheating), (c) properly designing the gate and runner, and (d) avoiding too large L/t ratio of the casting.

Fig. 12.12 Misrun casting
defect

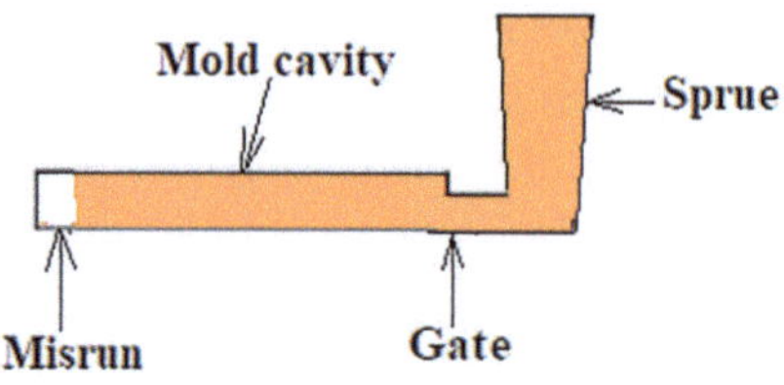

Fig. 12.13 Hot tear defect

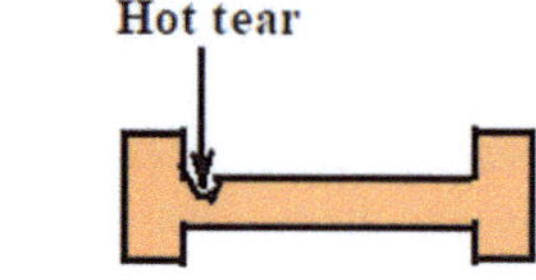

Fig. 12.14 Shrinkage
cavity defect in a casting

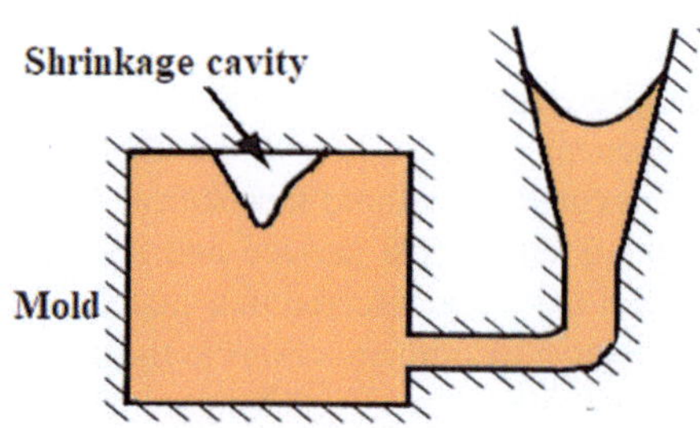

12.4.2.2   Hot Tearing (Hot Cracking)

A *hot tear* is a discontinuity that occurs during the solidification stage of a casting
process, when the material being *cast* is partly solid and partly liquid. It usually
appears as surface and sub-surface cracks (see Fig. 12.13). The reasons of the hot
teat defects may be one or more of the following: (a) excessively high temperature
of the melt, (b) increased metal contraction, and (c) incorrect gating-system design
and the casting design.

The hot tear defect can be avoided by the following techniques: (a) lowering the
metal temperature, (b) preventing metal contraction by using a sand mold of correct
hot strength, and (c) properly designing the gating system (e.g., correct sprue design,
correct runner design, correct gate design, etc.) (see Chap. 7, Sect. 7.4).

12.4.2.3   Shrinkage Cavity

A shrinkage cavity is a depression or ragged crevice in the center of a volume of
metal in a casting (see Fig. 12.14). It results from volume contraction during solidification. The shrinkage cavity defect is mainly caused by an incorrect riser design.
It can be prevented by ensuring that: (a) there is enough metal quantity in the riser,

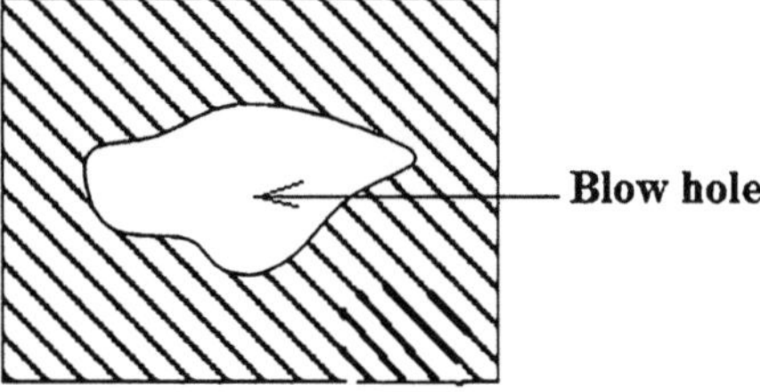

Fig. 12.15 Blow hole defect

and (b) the metal in the riser cools more slowly than the casting (see Chap. 7, Sect. 7.5).

12.4.2.4 Blow Hole

The *blow hole* or *sand blow* is a large gas bubble or cavity caused by mold gas during pouring (see Fig. 12.15). The reasons of the blow hole defect include low-permeability sand, poor venting, or high moisture content in the molding sand. The blow hole defect can be prevented by: (a) raising the permeability of sand by ensuring round-shaped sand particles, (b) proper venting, and (c) lowering the moisture content of the molding sand.

12.4.2.5 Pin Holes

Pin holes are small gas cavities at or slightly below the surface of the casting. They are mainly caused by the reaction between the metal oxides and the carbon of the melt. The other causes of pin holes include excessive hydrogen content in the melt, excessive nitrogen in the melt, and unfavorable gas atmosphere in the mold cavity (Khiani & Messer, 2017). Pin holes can be avoided by minimizing the contents of hydrogen, nitrogen, and sulfur in the melt.

In order to avoid pin holes in sand castings, it is important to keep the actual pressure within a flowing liquid in a sand mold above atmospheric pressure. If the pressure in the molten metal drops below 1 atmosphere, air can be drawn in or "aspirated" into the metal stream, thereby causing likelihood of defects within the casting. These defects could include the formation of metal oxides or gas porosity, a particularly serious problem in aluminum alloy castings which readily absorb hydrogen in the liquid state.

12.4.2.6 Metal Penetration

The *metal penetration* defect appears as a highly rough surface finish. It occurs when the liquid metal penetrates into the molding sand. This defect is caused by sand particles which are too coarse, lack of mold wash, or pouring temperatures that are too high. The *metal penetration* can be prevented by: (a) harder ramming of sand during the mold preparation, (b) using finer sand particles, and (c) reducing the pouring temperature.

12.4.2.7 Mismatch

A *mismatch* is the casting defect that occurs due to the misalignment of the lower and upper parts of the mold. This defect can be avoided by ensuring that the cope and drag of the mold are properly lined up before pouring the metal.

12.4.2.8 Porosity in Die Castings

The *porosity*, or *gas porosity*, is the formation of pores, holes, or pockets of air on the surface or within a die-cast part. The porosity occurs when air is trapped in the metal by the die casting machinery, often leaving gaps at the top of the die or filling a mold too slowly and having some solidification occur too soon. Recently, Dudek and co-researchers (2023) have conducted tests on castings made in the high-pressure die casting (HPDC) process and with the use of the vacuum system. The results of the tests showed that it is possible to increase the number of pressure die cavities and to obtain the required level of casting porosity, provided that the vacuum system is used. By using the vacuum technology, the amount of gas porosity in die castings (particularly aluminum die castings) can be significantly reduced from 4% at atmospheric pressure to 1% at a vacuum of 300 mbar (Dudek et al., 2023).

12.5 Worked Numerical Examples in Testing and QA

Example 12.1 Determining the Engineering Stress of a Metal at a Given Tensile Force

A tensile force of 1.5 kN is applied to a metal bar (original gauge length = 130 mm, original diameter = 2 mm). The final length at break is 155 mm. Determine the engineering stress.

Solution

$F = 1.5$ kN $= 1500$ N, $L_o = 130$ mm, $L_f = 155$ mm, $d_o = 2$ mm, $\sigma_{eng} = ?$

$$A_o = \frac{\pi}{4} d_o^2 = \frac{\pi}{4}(2)^2 = 3.1416\,\text{mm}^2$$

By using Eq. 12.1,

$$\sigma_{eng} = \frac{F}{A_o} = \frac{1500}{3.1416} = 477.464\,\text{MPa}$$

The engineering stress $= \sigma_{\text{eng}} = 477.464$ MPa.

Example 12.2 Determining the Engineering Strain at a Given Tensile Force

By using the data in Example 12.1, determine the engineering strain.

Solution

$L_0 = 130$ mm, $L_f = 155$ mm, $\varepsilon_{\text{eng}} = ?$

By using Eq. 12.2,

$$\varepsilon_{\text{eng}} = \frac{\Delta L}{L_0} = \frac{L_f - L_0}{L_0} = \frac{155 - 130}{130} = 0.192$$

The engineering strain $= 0.192$.

Example 12.3 Determining the Final Cross-Sectional Area of a Tensile Test Specimen

By using the data in Examples 12.1 and 12.2, determine the final cross-sectional area of the tensile-test specimen.

Solution

$A_0 = 3.1416$ mm^2, $L_0 = 130$ mm, $L_f = 155$ mm, $A_f = ?$

By using the constant volume relationship:

$$A_f L_f = A_0 L_0$$

$$A_f (155) = 3.1416 \times 130$$

$$A_f = 2.635\,\text{mm}^2$$

Example 12.4 Determining the Young's Modulus of a Metal

A tensile force of 2.5 kN is applied to a metal bar that is 145 mm long with a diameter of 1.8 mm. The length, at the proportional limit, is 147 mm. Determine the Young's modulus of the bar's material.

Solution

$F = 2.5$ kN $= 2500$ N, $l_0 = 145$ mm, $d_0 = 1.8$ mm, $l = 147$ mm, $E = ?$

$$A_0 = \frac{\pi}{4} d_0^2 = \frac{\pi}{4} 1.8^2 = 2.5447\,\text{mm}^2$$

$$\delta l = l - l_0 = 147 - 145 = 2\,\text{mm}$$

By using Eq. 12.5,

$$\text{Young's modulus} = E = \frac{F l_0}{A_0 (\delta l)} = \frac{2500 \times 145}{2.5447 \times 2} = 71{,}226.47\,\text{MPa} = 71.2265\,\text{GPa}$$

The Young's modulus of the material = E = 71.2265 GPa.

Example 12.5 Determining the Tensile Force Within the Proportional Limit
A tensile force is applied to an aluminum bar (original gage length = 170 mm, original diameter = 2.5 mm). The length of the rod at the proportional limit is 180 mm. Determine the tensile force.

Solution
Original gage length = l_o = 170 mm, d_o = 2.5 mm, l = 180 mm, F =?
   With reference to Table 12.1, we obtain: E = 70 GPa = 70,000 × 10^6 Pa = 70,000 MPa

$$A_o = \frac{\pi}{4} d_o^2 == \frac{\pi}{4} 2.5^2 = 4.91 \,\text{mm}^2$$

$$\delta l = l - l_o = 180 - 170 = 10 \,\text{mm}$$

By using the modified form of Eq. 12.5,

$$F = \frac{EA_o (\delta l)}{l_o} = \frac{70,000 \times 4.91 \times 10}{170} = 20,217.65 \,\text{N} = 20.217 \,\text{kN}$$

The tensile force = 20.217 kN.

Example 12.6 Determining the Proportional Limit and Elastic Strain Using the σ-ε Curve
Refer to the stress-strain curve for AISI-1020 steel (Fig. 12.4). Determine the (a) proportional limit, and (b) the elastic strain corresponding to the proportional limit for the steel.

Solution
The proportional limit and the corresponding elastic strain are marked in Fig. 12.16.

(a) By reference to Fig. 12.16, the proportional limit = stress = σ = 370 MPa.
(b) The elastic strain corresponding to the proportional limit is 0.0018 (see Fig. 12.16).

Example 12.7 Determining the Young's Modulus Using the Stress-Strain Curve
By using the data in Example 12.6, determine the Young's modulus for the steel.

Solution
σ = 370 MPa, ε_e = 0.0018, E =?
   By using the modified form of Eq. 12.3,

$$E = \frac{\sigma}{\varepsilon_e} = \frac{370}{0.0018} = 205.55 \times 10^3 \,\text{MPa} = 205.55 \,\text{GPa}$$

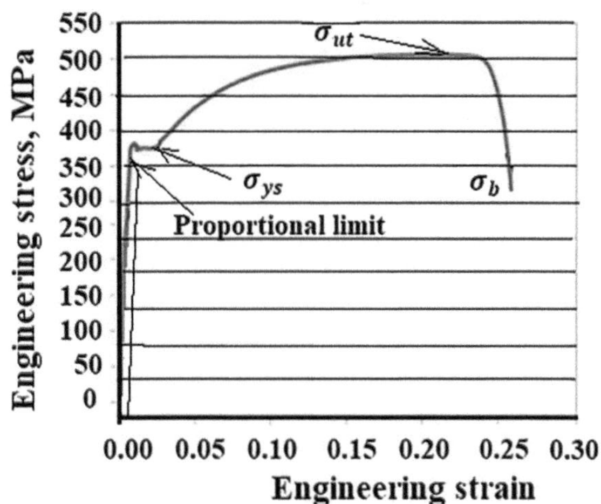

Fig. 12.16 The marked tensile stress-strain curve for AISI-1020 steel for Example 12.6

The Young's modulus of AISI-1020 steel = 205.55 GPa.

Example 12.8 Determining the Tensile Strength Using the Stress-Strain Curve for Steel

By using Fig. 12.4, determine the ultimate tensile strength of AISI-1020 steel.

Solution

With reference to Fig. 12.16, the maximum stress during the tensile test is 503 MPa.
Hence, the ultimate tensile strength of AISI-steel = σ_{ut} = 503 MPa.

Example 12.9 Determining the Yield Strength Using the Stress-Strain Curve for Steel

By using Fig. 12.4, determine the yield strength of AISI-1020 steel.

Solution

With reference to Fig. 12.16, σ_{ys} = 380 MPa.
Hence, the yield strength of AISI-1020 steel is 380 MPa.

Example 12.10 Determining the Breaking Strength Using the Stress-Strain Curve for Steel

By using Fig. 12.4, determine the breaking strength of AISI-1020 steel.

Solution

With reference to Fig. 12.16, σ_b = 320 MPa.
The breaking strength of AISI-1020 steel = 320 MPa.

Example 12.11 Calculating the Yield Strength Using the Tensile Test Data

A tensile test is conducted for a cylindrical specimen of stainless steel. The test specimen's original gage length is 50.8 mm and its original diameter is 12.8 mm. The force at the yield point is 93 kN. The maximum force, during the tensile test, is 160 kN, and the force at break is 125 kN. The final gage length at the break is 57 mm. Calculate the yield strength of the steel.

Solution
$F_y = 93$ kN $= 93{,}000$ N, $L_o = 50.8$ mm, $d_o = 12.8$ mm, $L_f = 57$ mm, $\sigma_{ys} =?$

$$A_o = \frac{\pi}{4}d_o^2 = 0.7854 \times 12.8^2 = 128.68\,\text{mm}^2$$

By using Eq. 12.6,

$$\sigma_{ys} = \frac{F_y}{A_o} = \frac{93{,}000}{128.68} = 722.72\,\text{N}/\text{mm}^2 = 722.72\,\text{MPa}$$

The yield strength of the steel $= 722.72$ MPa.

Example 12.12 Calculating the Ultimate Tensile Strength Using the Tensile Test Data
By using the data in Example 12.11, calculate the ultimate tensile strength of the steel.

Solution
$F_{max} = 160$ kN $= 160{,}000$ N, $A_o = 128.68$ mm^2, $\sigma_{ut} =?$
By using Eq. 12.7,

$$\sigma_{ut} = \frac{F_{max}}{A_o} = \frac{160{,}000}{128.68} = 1243.44\,\text{MPa}$$

The ultimate tensile strength of the steel $= 1243.44$ MPa.

Example 12.13 Calculating the Breaking Strength Using the Tensile Test Data
By using the data in Example 12.11, calculate the breaking strength of the steel.

Solution
$F_b = 125$ kN $= 125{,}000$ N, $A_o = 128.68$ mm^2, $\sigma_b =?$
By using Eq. 12.8,

$$\sigma_b = \frac{F_b}{A_o} = \frac{125{,}000}{128.68} = 971.42\,\text{MPa}$$

The breaking strength of the steel $= 971.42$ MPa.

Example 12.14 Calculating the Percent Elongation for a Metal
By using the data in Example 12.11, calculate the percent elongation in the steel.

Solution
$L_o = 50.8$ mm, $d_o = 12.8$ mm, $L_f = 57$ mm, % elongation $=?$
By using Eq. 12.9,

$$\%\text{Elongation} = \frac{L_f - L_o}{L_o} \times 100 = \frac{57 - 50.8}{50.8} \times 100 = 12.2$$

% Elongation in the steel = 12.2.

Example 12.15 Calculating the Percent Reduction in Area

By using the data in Example 12.11, calculate the percent reduction in area in the steel.

Solution

$A_o = 128.68$ mm^2, $L_o = 50.8$ mm, $L_f = 57$ mm, $A_f =$?, % reduction in area =?

By using the constant volume relationship,

$$A_f L_f = A_o L_o$$

$$A_f (57) = 128.68 \times 50.8$$

$$A_f = 114.68 \, \text{mm}^2$$

By using Eq. 12.10,

$$\%RA = \frac{A_o - A_f}{A_o} \times 100 = \frac{128.68 - 114.68}{128.68} \times 100 = 10.88$$

The percent reduction in area in the steel = 10.88.

Example 12.16 Determining the Engineering Strain Using the Stress-Strain Curve

With reference to Fig. 12.4, determine the engineering strain in the AISI-1020 steel.

Solution

The engineering strain at fracture is marked in Fig. 12.17.

With reference to Fig. 12.17, the engineering strain at fracture in the AISI-1020 steel = 0.26.

Example 12.17 Determining the Percent Elongation Using the Stress-Strain Curve

By using the stress-strain curve (Fig. 12.17), determine the percent elongation in the AISI-1020 steel.

Solution

Engineering strain = 0.26, % elongation =?

By using Eq. 12.9,

$$\%\text{Elongation} = \frac{\Delta l}{l_o} \times 100 = (\text{engineering strain}) \times 100 = 0.26 \times 100 = 26$$

The percent elongation in the AISI-1020 steel = 26.

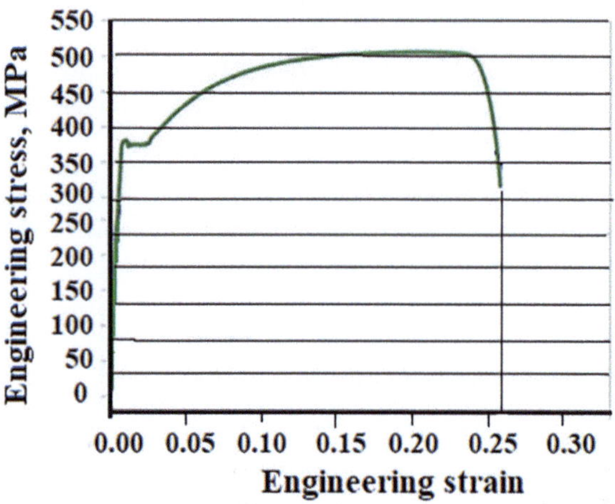

Fig. 12.17 The stress-strain curve showing the engineering strain at fracture (l_o = 50 mm)

Example 12.18 Determining the Final Length at Fracture Using the Stress-Strain Curve

By using the stress-strain curve (Fig. 12.17) in Example 12.17, determine the final length at fracture in the tensile test.

Solution

With reference to the figure caption of Fig. 12.17, l_o = 50 mm, % Elongation = 26.
By using Eq. 12.9,

$$\%\text{Elongation} = \frac{l_f - l_o}{l_o} \times 100$$

$$26 = \frac{l_f - 50}{50} \times 100 = 2 \times (l_f - 50)$$

$$l_f - 50 = 13$$

$$l_f = 63 \ \text{mm}$$

The final length at fracture = 63 mm.

Example 12.19 Determining the Yield Strength in Compression

Figure 12.18a shows the stress-strain curves for gray cast iron, as obtained by testing in a universal testing machine. Determine the yield strength in compression.

Solution

The yield point in compression is marked in Fig. 12.18b.
With reference to Fig. 12.18b, the yield strength in compression for gray cast iron = 320 MPa.

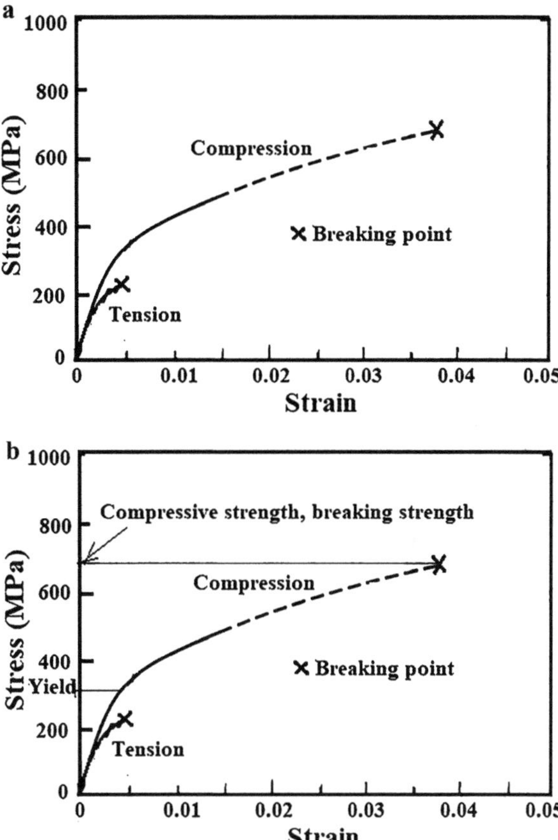

Fig. 12.18 (**a**) The stress-strain curve for gray cast iron. (**b**) The marked stress-strain curve for Example 12.19

Example 12.20 Determining the Compressive Strength of a Metal

By using the data in Example 12.19, determine the compressive strength of gray cast iron.

Solution

With reference to Fig. 12.18b, the compressive strength of gray cast iron = 700 MPa.

Example 12.21 Determining the Breaking Strength in Compression for a Metal

By using the data in Example 12.19, determine the breaking strength in compression for the tested cast iron.

Solution

With reference to Fig. 12.18b, the breaking strength in compression for the iron = 700 MPa.

Example 12.22 Determining the Brinell Hardness Number When the Applied Load Is Known

A load of 1000 kgf was applied by using a 10-mm diameter steel ball that resulted in the indentation diameter of 8 mm in a test material. Determine the BHN for the test material.

Solution

Data: $F = 1000$ kgf, $D = 10$ mm, $D_i = 8$ mm, BHN =?

By using Eq. 12.11,

$$\text{BHN} = \frac{F}{\frac{\pi}{2}\left(D - \sqrt{D^2 - D_i^2}\right)} = \frac{1000}{\frac{\pi}{2}\left(10 - \sqrt{10^2 - 8^2}\right)} = \frac{1000}{\frac{\pi}{2}(10-6)} = \frac{1000}{6.2832} = 159.15 \cong 159$$

Example 12.23 Specifying a BHN Indicating the Hardness Test Conditions

A Brinell hardness number of 300 was obtained using a 10-mm diameter hardened steel ball with a 3000 kgf load applied for 10 s. Specify the BHN indicating the test condition.

Solution

The well-defined BHN is given as follows: 300 HB 10/3000/10.

Example 12.24 Determining the Ultimate Tensile Strength of Steel When BHN Is Known

The BHN of test sample of carbon steel was determined to be 200. Estimate the ultimate tensile strength of the steel. Verify the answer by comparing with the data in Table 12.1.

Solution

BHN = 200, σ_{ut} =?

By using Eq. 12.12,

$$\sigma_{ut} = 4.35 * (\text{BHN}) = 4.35 \times 200 = 870\,\text{MPa}$$

The ultimate tensile strength of 870 MPa is well within the range of 350–2000 MPa.

Example 12.25 Determining the Rockwell Hardness Number

The Rockwell hardness testing was conducted on a metal sample using a steel ball indenter (scale: C). The permanent increase in the depth of penetration was recorded as 0.14 mm. Determine the Rockwell hardness number.

Solution

$R = 130$ (for steel ball indenter), $s = 0.002$ mm, $h = 0.14$ mm, HR =?

By using Eq. 12.3,

$$\text{HR} = R - \frac{h}{s} = 130 - \frac{0.14}{0.002} = 130 - 70 = 60$$

The Rockwell hardness number is 60HRC.

Example 12.26 Calculating the Arithmetic Mean of the Diagonals in Vickers Hardness Test

A load of 80 kgf is applied for 10 s using the indenter of a Vickers hardness tester. The two diagonals of the indentation on the surface of the test material are 0.8 and 1 mm, respectively. Calculate the arithmetic mean of the diagonals.

Solution

$d_1 = 0.8$ mm, $d_2 = 1.0$, $d =$?

$$d = \frac{d_1 + d_2}{2} = \frac{0.8 + 1}{2} = 0.9\,\text{mm}$$

Example 12.27 Determining the Vickers Hardness Number

By using the data in Example 12.26, determine and completely specify the Vickers hardness number for the test material.

Solution

$F = 80$ kgf, $d = 0.9$ mm, VHN $=$?

By using Eq. 12.14,

$$\text{VHN} = \frac{1.854 * F}{d^2} = \frac{1.854 \times 80}{0.9^2} = 183.11$$

The Vickers hardness number of the test material is 183 VHN/80.

Example 12.28 Determining the Knoop Hardness Number

A load of 400 gf was applied on a test material by use of a Knoop indenter. The longest diagonal of the indentation is measured to be 85 µm. Determine the Knoop hardness number.

Solution

$P = 400$ gf $= 0.4$ kgf, $L = 85$ µm $= 0.085$ mm, KHN $=$?

By using Eq. 12.15,

$$\text{Knoop hardness number} = \text{KHN} = \frac{14.23\,P}{L^2} = \frac{14.23 \times 0.4}{0.085^2} = \frac{5.692}{0.007225}$$

$$= 787.82 \cong 788$$

Example 12.29 Determining the Impact Energy of a Material

A Charpy impact test was conducted by using a heavy pendulum of 35 kgf that is released from a height of 30 cm. The pendulum struck the specimen on its downward swing thereby fracturing it. The height at the end of swing is 3 cm. Determine the impact energy of the test material.

Solution

$m = 35$ kgf, $g = 9.81$ m/s², $h - h' = 30 - 3 = 27$ cm $= 0.27$ m, $U_i = ?$

By using Eq. 12.16,

$$U_i = mg(h - h') = 35 \times 9.81 \times 0.27 = 92.7 \, \text{J}$$

Example 12.30 Computing the Mean Stress Acting on a Component

The bending stresses acting on a component were determined to be ±80 MPa. Determine the mean stress acting on the component.

Solution

$\sigma_{max} = 80$ MPa, $\sigma_{min} = -80$ MPa, $\sigma_m = ?$

By using Eq. 12.17,

$$\text{Mean stress} = \sigma_m = \frac{\sigma_{max} + \sigma_{min}}{2} = \frac{80 + (-78)}{2} = 0$$

Example 12.31 Calculating the Stress Range Acting on a Component

By using the data in Example 12.30, calculate the stress range acting on the component.

Solution

By using Eq. 12.18

$$\text{Stress range} = \sigma_r = \sigma_{max} - \sigma_{min} = 80 - (-80) = 80 + 80 = 160 \, \text{MPa}$$

Example 12.32 Calculating the Stress Amplitude

By using the data in Example 12.31, calculate the stress amplitude.

Solution

$\sigma_r = 160$ MPa, $\sigma_a = ?$

By using Eq. 12.19,

$$\text{Stress amplitude} = \sigma_a = \frac{\sigma_r}{2} = \frac{160}{2} = 80 \, \text{MPa}$$

Example 12.33 Determining the Stress Ratio

By using the data in Example 12.30, determine the stress ratio.

Solution

By using Eq. 12.20,

$\sigma_{max} = 80$ MPa, $\sigma_{min} = -80$ MPa, $R_s = ?$

$$\text{Stress ratio} = R_{s} = \frac{\sigma_{min}}{\sigma_{max}} = \frac{-80\,\text{MPa}}{80\,\text{MPa}} = -1$$

Example 12.34 Determining the Amplitude Ratio

By using the data in Example 12.30, determine the amplitude ratio.

Solution

$\sigma_{a} = 80$ MPa, $\sigma_{m} = 0$, $R_{a} = ?$

   By using Eq. 12.21,

$$\text{Amplitude Ratio} = R_{a} = \frac{\sigma_{a}}{\sigma_{m}} = \frac{80}{0} = \infty$$

Example 12.35 Determining the Fatigue Limit of the Test Material

A rotating steel shaft is subjected to a stress range of 350 MPa. The tensile strength of the shaft steel is 450 MPa. Determine the fatigue limit of the test material.

Solution

$\sigma_{ut} = 450$ MPa, $S_{e} = ?$

   By using Eq. 12.22,

$$S_{e} = \frac{\sigma_{ut}}{2} = \frac{450}{2} = 225\,\text{MPa}$$

Example 12.36 Determining the Mean Stress Using Goodman's Law

By using the data in Example 12.35, determine the maximum mean stress the steel shaft is able to withstand.

Solution

$\sigma_{ut} = 450$ MPa, $S_{e} = 225$ MPa, $\sigma_{r} = 350$ MPa, $\sigma_{m} = ?$

$$\sigma_{a} = \frac{\sigma_{r}}{2} = \frac{350}{2} = 175\,\text{MPa}$$

   By using Goodman's law,

$$\frac{\sigma_{a}}{S_{e}} + \frac{\sigma_{m}}{\sigma_{ut}} = 1$$

$$\frac{175}{225} + \frac{\sigma_{m}}{450} = 1$$

$$\frac{\sigma_{m}}{450} = 1 - 0.777 = 0.2223$$

$$\text{Mean stress} = \sigma_m = 0.2223 \times 450 = 100\,\text{MPa}$$

Example 12.37 Determining the Incident Angle Using Snell's Law in UT

The sound velocity of a longitudinal wave in Lucite is 0.106 in/μs, and the sound velocity of a shear wave in steel is 0.128 in/μs. Determine the incident angle that will produce a 75° refracted shear wave in steel using a Lucite wedge.

Solution

$v_1 = 0.106$ in/μs, $v_2 = 0.128$ in/μs, $\theta_2 = 75°$, $\theta_1 = ?$

By using Eq. 12.24,

$$\frac{\sin\theta_1}{\sin 75°} = \frac{0.106}{0.128}$$

$$\frac{\sin\theta_1}{0.9659} = \frac{0.106}{0.128} = 0.8281$$

$$\sin\theta_1 = 0.8281 \times 0.9659 = 0.79986$$

$$\theta_1 = 53°$$

Example 12.38 Determining the Near Field in Ultrasonic Testing

The short-pulse waves with a frequency of 7 MHz are emitted from a 0.38-in diameter transducer to inspect a component made of brass. Determine the end of the near field. The sound velocity in brass is 1.685×10^5 in/s.

Solution

$D = 0.38$ in., $f = 7$ MHz $= 7 \times 10^6$ Hz, $v = 1.685 \times 10^5$ in/s, $N = ?$

By using Eq. 12.25,

$$N = \frac{D^2 * f}{4v} = \frac{0.38^2 \times 7 \times 10^6}{4 \times 1.685 \times 10^5} = \frac{10.108}{6.74} = 1.4997\,\text{in}$$

The end of the near field $= 1.4997$ in $\cong 1.5$ in.

Example 12.39 Determining the Beam Spread Angle in Radiographic NDT

By using the data in Example 12.38, determine the beam spread angle.

Solution

$D = 0.38$ in., $f = 7$ MHz $= 7 \times 10^6$ Hz, $v = 1.685 \times 10^5$ in/s, $\theta = ?$

By using Eq. 12.26,

$$\sin\theta = 1.2 * \left(\frac{v}{D * f}\right) = \left(\frac{1.2 \times 1.685 \times 10^5}{0.38 \times 7 \times 10^6}\right) = \frac{0.2022}{2.66} = 0.076$$

$$\theta = 4.3°$$

Example 12.40 Determining the Radiation Intensity in Radiographic NDT Examination

In a radiographic NDT examination, the intensity of the received dose of radiation, at a distance of 2 cm from the radiation source, is 120 mR/h. Determine the radiation intensity at a distance of 2 m from the source.

Solution

$I_1 = 120$ mR/h, $d_1 = 2$ cm, $d_2 = 2$ m $= 200$ cm, $I_2 =$?

By using Eq. 12.27,

$$120 \times 2^2 = I_2 \times 200^2$$

$$480 = 40,000\, I_2$$

$$I_2 = \frac{480}{40,000} = 0.012\,\text{mR}/\text{h}$$

Questions and Problems

12.1. **MCQs**: Encircle the most appropriate answer for each of the following statements.

(1) Which mechanical test involves indentation in the test material?
   (a) Fatigue test, (b) impact test, (c) hardness test, (d) tensile test

(2) Which mechanical test is the best for determining yield strength?
   (a) Tensile test, (b) impact test, (c) fatigue test, (d) hardness test

(3) Which hardness test involves the use of rhombic-based pyramidal diamond indenter?
   (a) Brinell test, (b) Knoop test, (c) Rockwell test, (d) Vickers test

(4) Which hardness test allows us to use hardened steel ball indenter?
   (a) Brinell test, (b) Vickers test, (c) Knoop test, (d) Rockwell test

(5) The maximum stress during tensile testing is called ____________.
   (a) Breaking strength, (b) ultimate tensile strength, (c) fatigue strength, (d) yield strength

(6) The increase in stress beyond the yield stress to the maximum stress is called _____.
   (a) Strain hardening, (b) yield stress, (c) precipitation hardening, (d) surface hardening

(7) Fatigue failure occurs due to the application of __________________.
   (a) Impact loading, (b) gradual loading, (c) cyclic loading, (d) high strain loading

(8) One-half of the stress range, in cyclic loading, is called ____________.
   (a) Mean stress, (b) maximum stress, (c) stress ratio, (d) stress amplitude

(9) Which type of failure usually occurs in gears?

   (a) Brittle failure, (b) fatigue failure, (c) ductile failure, (d) none.

(10) Which type of NDT involves capillarity principle during testing?

   (a) Magnetic particle test, (b) ultrasonic test, (c) dye-penetrant test, (d) radiography

(11) Which type of NDT is the simplest for detecting surface flaws?

   (a) Radiography, (b) ultrasonic test, (c) dye penetrant test, (d) magnetic particle test

(12) Which type of NDT involves the use of Snell's law?

   (a) Magnetic particle test, (b) ultrasonic test, (c) dye penetrant test, (d) radiography

(13) Which type of NDT is the best for detecting both small and large internal defects?

   (a) Magnetic particle test, (b) ultrasonic test, (c) dye penetrant test, (d) radiography

(14) Which defect is most likely to result by too low melt temperature and too large length-to-thickness ratio (L/t) of the casting?

   (a) Blow hole, (b) misrun, (c) shrinkage cavity, (d) porosity

(15) Which defect is most likely to result by high hydrogen, nitrogen, and sulfur in the melt?

   (a) Blow hole, (b) misrun, (c) shrinkage cavity, (d) pin holes

(16) Which casting defect is most likely to result by incorrect riser design?

   (a) Blow hole, (b) misrun, (c) shrinkage cavity, (d) porosity

(17) Which defect is most likely to result by high moisture content and improper venting?

   (a) Blow hole, (b) misrun, (c) shrinkage cavity, (d) pin holes

(18) Which defect may result due to very high melt temperature and incorrect gating system?

   (a) Blow hole, (b) misrun, (c) shrinkage cavity, (d) hot tears

(19) Which defect is the most common in aluminum die casting?

   (a) Blow hole, (b) misrun, (c) shrinkage cavity, (d) porosity

(20) What is the frequency range of the waves in ultrasonic testing?

   (a) 0.1–15 MHz, (b) 50–500 MHz, (c) 0.1–15 kHz, (d) 50–500 kHz

12.2. Draw a sketch of tensile-test specimen and briefly explain tensile testing procedure.

12.3. Describe the Vickers hardness test with the aid of a sketch.

12.4. List the various NDT techniques. Why is it important to carry out NDT of castings?

12.5. Explain ultrasonic test with the aid of a sketch.

12.6. (a) List seven casting defects.

   (b) How will you avoid the following casting defects: hot tears, misrun, and blow holes?

   (c) What recent technology has been developed to overcome porosity is aluminum die castings?

P-12.7. A tensile test is conducted for a cylindrical specimen of steel. The test specimen's original gage length is 50.8 mm and its original diameter is 12.8 mm. The force at the yield point is 98 kN. The maximum force, during the tensile test, is 180 kN, and the force at break is 135 kN. The final gage length at the break is 60 mm. Determine the following mechanical properties of the steel: (a) yield strength, (b) ultimate tensile strength, (c) breaking strength, (d) % elongation, and (e) % reduction in area.

P-12.8. The BHN of test sample of steel was determined to be 400. Estimate the ultimate tensile strength of the steel.

P-12.9. A load of 1500 kgf was applied by using a 10-mm diameter steel ball that resulted in the indentation diameter of 9 mm in a test material. Determine the BHN of the test material.

P-12.10. The Rockwell hardness testing was conducted on a metal sample using a steel ball indenter (C scale). The permanent increase in the depth of penetration was 0.16 mm. Determine the Rockwell hardness number.

P-12.11. A load of 70 kgf is applied for 8 s using the indenter of a Vickers hardness tester. The two diagonals of the indentation on the surface of the test material are 0.6 and 0.8 mm, respectively. Determine the Vickers hardness number.

P-12.12. A load of 700 gf was applied on a test material by use of a Knoop indenter. The longest diagonal of the indentation is measured to be 100 μm. Determine the KHN.

P-12.13. A Charpy impact test was conducted by using a pendulum of 30 kgf that is released from a height of 25 cm. The pendulum struck the specimen thereby fracturing it. The height at the end of swing is 2.5 cm. Determine the impact energy of the test material.

P-12.14. A rotating steel shaft is subjected to a stress range of 350 MPa. The tensile strength of the shaft steel is 500 MPa. Determine the fatigue limit of the test material.

P-12.15. The bending stresses acting on a component were determined to be ±70 MPa. Determine the: (a) mean stress, (b) stress range, (c) stress amplitude, and (d) stress ratio.

P-12.16. The velocities of sound for a longitudinal wave in Lucite and a shear wave in steel are 0.106 and 0.128 in/μs, respectively. Determine the incident angle that will produce a 70° refracted shear wave in steel using a Lucite wedge.

P-12.17. The short-pulse waves with a frequency of 8 MHz are emitted from a 0.4-in diameter transducer to inspect a component made of brass. Determine the end of the near field. The sound velocity in brass is 1.685 × 10^5 in/s.

References

ASTM Committee. (2013). *Designation: E8/E8M–13a: Standard test methods for tensile testing of metallic materials.* https://www.galvanizeit.com/uploads/ASTM-E-8-yr-13.pdf

CNE-ISU. (2024). *Non-destructive evaluation techniques, center of non-destructive evaluation.* Iowa State University. Retrieved November 29, 2024, from https://www.nde-ed.org/NDETechniques/Ultrasonics/ultrasonicFormula.xhtml

Dudek, P., Białoń, J., Piwowońska, J., Walczak, W., & Wrzała, K. (2023). The impact on the cost of making high pressure die castings with multi-cavity die and vacuum assistance. *Vacuum, 210,* 111859.

Goodrich, G. (2008). *Casting defects handbook: Iron & steel.* American Foundry Society.

Huda, Z. (2020). *Metallurgy for physicists and engineers.* CRC Press.

Huda, Z. (2022). *Mechanical behavior of materials.* Springer Publishing.

Huda, Z., Ajani, M. H., & Ahmed, M. S. (2020). Fatigue behaviors of two notched cutting-tool materials: High speed steel and cemented carbide. *Material Testing, 62*(3), 265–270.

Khiani, G., & Messer, B. (2017, December). Pinhole defects in a casting: Causes and corrective action. *Valve World Americas.* https://gapvinc.wordpress.com/wp-content/uploads/2019/03/page7-9columngobinddec.pdf

Pelleg, J. (2013). *Mechanical properties of materials.* Springer Publishing.

Chapter 13
Safety and Lean Manufacturing in Foundry

Abbreviations

DfM   Design for manufacturability
INCMA   Indiana Cast Metal Association
JSHIRA   Job Safety Hazard Identification and Risk Assessment
OSHA   Occupational Safety and Health Administration
PPE   Personal protective equipment

13.1 Foundry Safety

13.1.1 Why Is Foundry Safety Important?

The safety, in a foundry, must be given the utmost attention among all facilities that perform metal casting, since there are many safety and health issues to be aware of when performing a casting operation. The most important foundry-safety concern is burning, as employees are working with extremely hot molten metals. If a worker comes in contact with the hot material or equipment, they will be severely injured. In addition to burns, foundry-safety programs should also address dust inhalation, heat stress, vision protection, and sparks (in the case of a cupola).

In order to ensure/improve safety and protect the foundry company from liability, the following ten safety precautions and measures should be implemented in the casting facilities: (1) conducting a thorough hazard assessment, (2) ensuring workers wear personal protective equipment (PPE), (3) compliance with environmental regulations, (4) recognition with signs of heat stress, (5) keeping moisture out, (6) inspecting machinery before use, (7) reducing dust exposure, (8) monitoring noise levels, (9) preventing and minimizing harmful vibrations, and (10) regular training of employees (PumpWorks Castings, 2004). These safety measures are explained in the following subsections.

Z. Huda, *Metal Casting Engineering*, Mechanical Engineering Series,
https://doi.org/10.1007/978-3-031-84620-5_13

273

13.1.2   Hazard Assessment in Foundry

A thorough hazard assessment, in a foundry, will help to identify all health and safety issues in metal casting. Although all foundries encounter some of the same casting safety issues, each operation is unique depending on the casting equipment and procedure. The thorough assessment will, therefore, help a foundry facility to determine which hazard poses the most considerable risk to the organization. In particular, health hazards in investment casting must be properly addressed, since this process involves vaporization of wax which may cause inhalation issues.

Recently, Rajkumar and co-researchers have conducted job safety hazard identification and risk analysis in a foundry. They have reported that *JSHIRA* (Job Safety Hazard Identification and Risk Assessment) is an effective risk-assessment management tool for ensuring safety in foundries. The *JSHIRA* method consists of various sequences of procedures that are implemented in the foundry division workplace (Rajkumar et al., 2021)

13.1.3   Personal Protective Equipment (PPE)

It is important that workers wear personal protective equipment (PPE) to avoid hazards and to ensure a safe and productive workplace. The OSHA (Occupational Safety and Health Administration) regulations for PPE in foundries require workers to wear protective clothing made of protective materials (e.g., leather, aluminized fabric, etc.). Additionally, workers should be asked to wear protective equipment (such as respirators, eye protection, face protection, etc.) to protect against hazards like molten metal (see Fig. 13.1). The right clothing and equipment can reduce a worker's risk of being burnt, inhaling dust, or incurring eye damage.

13.1.4   Compliance with Environmental Regulations

The company (say Foundry Corporation *X*) must take care the health and safety of its employees; additionally, the protection of the natural environment is a critical concern in the operation of a foundry's business. It should, therefore, be the policy of Foundry Corporation *X* to: (a) actively pursue process innovation to reduce and eliminate waste from its operations and prevent environmental pollution; (b) routinely review and assess its operations for the purpose of making continual improvements in areas of health, safety and environmental concern; and (c) comply with all applicable laws, regulations, and standards in its product development, manufacturing, marketing, and distribution activities (INCMA, 2004).

Fig. 13.1 Personal protective equipment (PPE)

13.1.5   Managing Heat Stress In Foundry

Heat stress, heat exhaustion, and heat stroke are big heat-related risks in a foundry. A high-heat environment can cause severe health issues for employees. The foundry employees should, therefore, be trained to recognize the signs of heat stress so that they can look after one another, provide the basic first aid, and notify management, accordingly.

13.1.6   Keeping Moisture Out

A high moisture level, in a foundry environment, causes explosive bursts of steam that scald employees. It is, therefore, important that facility managers make every effort to minimize moisture on the foundry floor. The metal scrap, meant to be charged in a melting furnace, should be warmed before melting it. Additionally, the molding sand should be stored in moisture-tight areas.

13.1.7   Inspecting Machinery

A faulty machinery (such as those with loose fastenings) can cause injury to the operators. The equipment should, therefore, be thoroughly inspected before it is used. This inspection will ensure that the equipment is functioning correctly and free from risks. The inspection of machinery allows facility managers to detect potential issues sooner. Besides safety, the inspection will also extend the lifespan of equipment, prevent casting defects, and reduce unexpected production delays.

13.1.8 Reducing Dust Exposure

A foundry worker may inhale dust, which may cause several health issues. For example, the additional processing after removal of castings from molds may involve cleaning, grinding, and sand basting, and these operations may cause exposure to dust. A foundry manager can minimize these risks by proactively reducing the dust produced and installing quality ventilation equipment. The workers should also be provided with personal protective equipment (PPE), such as respirators. As a whole, these safety measures will help foundries address many health and safety issues in metal casting.

13.1.9 Monitoring Noise Level

Excessive noise levels can damage workers' hearing abilities. Additionally, noise pollution makes it difficult for employees to communicate; thus, they will not be able to collaborate as effectively and might find it hard to warn one another of hazards. It is, therefore, important that foundry managers monitor noise levels within the workplace. In the case of high noise levels, noise-dampening equipment should be installed, as doing so will improve safety and make it easier for employees to communicate.

13.1.10 Preventing and Minimizing Harmful Vibrations

The machinery, during its operation, should run with minimal noise and harmful vibrations. Noisy equipment indicates harmful vibrations and a risk of damage and safety hazard to the machinery operators. All harmful vibrations must be controlled and minimized. This would significantly improve safety in several ways. Firstly, reducing vibration minimizes equipment damage, which leads to a lower risk of catastrophic machinery failure. Secondly, less vibration also improves employees' communication on the shop floor.

13.1.11 Training of Employees

In many manufacturing situations, employees have to repair foundry equipment. This situation may result in ignorance of safety precautions. This author has seen a technician whose face was burnt because he was not cautious about safety precautions during the repair of a furnace. In general, employees can become complacent regarding health and safety issues during a casting process. This is why foundry managers must reduce this complacency by providing staff with regular/scheduled training on safety.

13.2 Lean Manufacturing

13.2.1 Lean Manufacturing: Advantages and Principles

Lean manufacturing is a production process based on a methodology of minimizing waste and maximizing productivity within a manufacturing set up. The lean manufacturing process focuses on ways to improve quality while reducing waste. Companies that successfully implement lean manufacturing methodology often find their costs dropping over time. There are four major advantages of lean manufacturing: (a) waste minimization, (b) cost reduction, (c) meet promised dates, and (d) quality improvement.

Lean manufacturing is based on the following five basic principles: (1) defining value, (2) mapping the value stream, (3) creating flow, (4) using a pull system, and (5) pursuing perfection (see Fig. 13.2). The scope of this book does not allow the author to give a detailed explanation of the five principles. The reader may refer to the relevant literature (e.g. Dailey, 2003) for a detailed discussion on lean manufacturing.

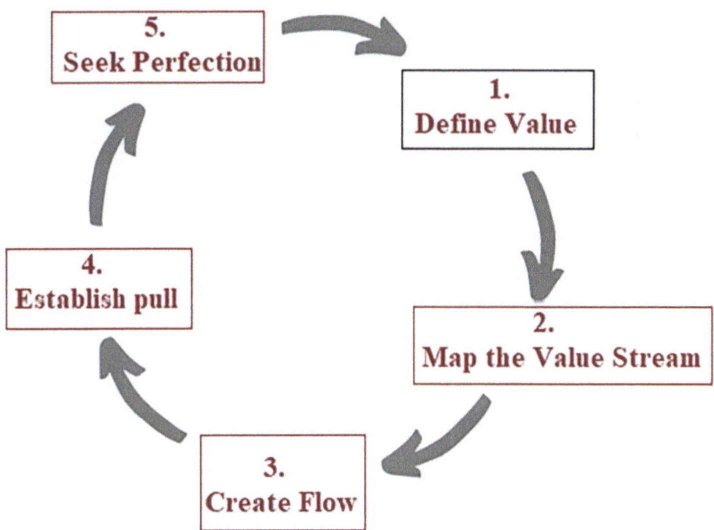

Fig. 13.2 The five principles of lean manufacturing

13.2.2   Lean Manufacturing in Foundry

The term *lean production* means a continuous effort toward production excellence by minimizing any kind of waste. There are seven types of waste in manufacturing/foundries: (1) inventory, (2) waiting, (3) defects, (4) over-production, (5) motion, (6) transportation, and (7) over-processing. Each one of this kind of waste is a useless activity that uses resources, including time, without any benefit (Calogero et al., 2015; Longo et al., 2019).

In order to reduce waste in foundries, the following lean-manufacturing approaches are useful: (a) castability, (b) bottleneck solutions, and (c) optimizing cycle time. These lean techniques are explained in the following paragraphs.

Castability   The molder is looking for a casting design that allows small variations in the pouring speed and metal temperature as well as being easy to eject. A casting design with thin walls, sharp edges, or deep pockets indicate poor castability resulting in waste, hence these undesirable features in the casting should be avoided. A good casting design will help reduce the costs of scrap and lost time, which can contribute to the overall cost reduction of each casting.

Bottleneck Solutions   Bottlenecks occur in metal casting when one aspect of the process takes longer, or is less efficient, than other aspects. This situation results in a build-up of cast parts in an unfinished stage, and a backlog of cast parts stuck behind the buildup. By using lean methodology, the bottleneck problem can be overcome by optimizing the workflow in the foundry. Bottlenecks also occur in the cleaning room of foundries. Lean manufacturing practices ensure the maximum efficiency and throughput by eliminating waste of time in the cleaning room. The size of the casting, as well as the time and the type of cleaning, will contribute to the bottleneck solutions.

Optimizing Cycle Time   A reduction in cycle time, in general, significantly reduces waste; however, in some cases, extending a cycle time helps to reduce the number of employees needed to produce the casting. For example, increasing the cooling time of the casting would allow a secondary process to be moved into the molding function. Thus, consolidation of functions may necessitate fewer employees on a job, freeing up operators to work on other tasks

13.2.3   Lean Manufacturing in Sand Casting: A Case Study

There are many different types of metal casting processes (see Chap. 1); however, the most common processes are based on either sand casting or die casting. For example, sand casting entails the following primary steps: mold preparation, melting the metal, pouring, and breaking open the mold after it has cooled. In order to implement lean manufacturing, the foundry manager/engineer should map out each step of operation, and look for inefficient aspects within each step. The engineer may create a flow chart for mapping out each step of the casting operation. In some cases, changing (improving) a foundry's physical layout (plant layout) could also help a company embrace lean manufacturing principles and reduce overall casting costs.

Recently, Saettaa and Caldarelli have reported a case study on the lean and green production methodology in a sand-casting foundry. They have analyzed the core making process in the Department of Core Shooting Machines of the foundry. Previously, the production cycle of the cores (with organic binder) involved the following four phases: (a) core production, (b) core warehousing, (c) core working in progress, and (4) use in the casting line. By implementing lean/green production methodology, the use of organic binder was replaced by inorganic binder, thereby eliminating one phase of the production cycle of the cores: the core warehousing (Saettaa & Caldarellia, 2023).

13.2.4   Lean Manufacturing in Die-Casting Foundry

Lean manufacturing in a die-casting (especially high-pressure die casting [HPDC]) facility can be achieved by considering the following aspects: (a) material selection and cost implications, (b) part complexity and design considerations, (c) production volume and economy, (d) tooling and die cost, and (e) reducing rejections by vacuum technology. These lean-manufacturing aspects are briefly explained in the following sections.

13.2.4.1   Material Selection for Lean Manufacturing in HPDC

The selection of material plays a significant role in achieving lean manufacturing goals, particularly in reducing the die-casting cost. Based on their availability and properties, different metals have varying costs. Additionally, material wastage, recyclability, and post-processing requirements also impact the overall cost. It is important to evaluate the specific requirements of the part and select a material that strikes a balance between cost-effectiveness and performance. The material utilization rate of die casting is generally in the range of 90–95%.

13.2.4.2   Part Complexity and Design for Manufacturability (*DfM*)

The complexity of part directly affects the die-casting cost. In particular, thin sections, intricate designs, undercuts, and tight tolerances increase the complexity of the mold and the production process. In order to optimize the part geometry, *design for manufacturability (DfM)* principles should be applied for cost-effective die casting. The implementation of *DfM* reduces the need for secondary operations and minimizes material usage.

13.2.4.3   Production Volume and Economy

The production volume has a significant impact on HPDC costs. The HPDC process is not economically feasible for small volumes of production. Due to the significant cost of the pressure die, the HPDC technology is profitable only in the case of large-scale production. This is why hot-chamber HPDC machines are designed with production volumes as high as 1000 units (castings) per hour (Herman & Zikmund, 2010).

13.2.4.4   Tooling and Die Cost

The cost of tooling and dies is an essential consideration in die casting. The overall cost is strongly influenced by tool and mold design complexity, size, and material selection. The tooling life and the maintenance costs should also be considered to ensure long-term cost-effectiveness.

13.2.4.5   Reducing Rejections in Die Casting by Vacuum Technology

The minimization of waste is directly related to the reduction of rejections in a foundry. Experiments in HPDC have shown that by increasing the number of pressure die cavities and by using the vacuum technology, the amount of gas-porosity defect in die castings (particularly aluminum die castings) can be significantly reduced from 4% at atmospheric pressure to 1% at a vacuum of 300 mbar (Dudek et al., 2023). This research finding significantly contributes to lean manufacturing in a die casting facility.

Questions

13.1. MCQS. Encircle the most appropriate answer(s) for each of the following statements

   (1) Which type of hazard needs the most attention in investment casting?
       (a) Burn hazard, (b) health (inhaling) hazard, (c) injury hazard, (d) hearing issue
   (2) Which technique is used to minimize waste in a casting's design?
       (a) Design for manufacturability, (b) design for manufacturing, (c) functional design
   (3) The situation of build-up of cast parts in an unfinished stage, and a backlog of cast parts stuck behind the buildup, is called ____________:
       (a) Inventory control, (b) lean manufacturing, (c) green production, (d) bottleneck

13.2. Why is safety important in a foundry?

13.3. List at least eight safety measures to be implemented in a foundry/industrial set-up.

13.4. List the seven types of wastes in manufacturing.

13.5. Draw a diagram showing the five principles of lean manufacturing.

13.6. What are the benefits of inspection of machinery before its use?

13.7. As a foundry engineer, how will you implement lean manufacturing in sand casting?

13.8. How does the use of vacuum technology and increase in die cavities contribute to lean manufacturing in die casting foundry?

References

Calogero, A., Longo, F., Nicoletti, L., Massei, M., & Solis A. (2015). Lean management practices to improve performances in operating rooms. In *4th International Workshop on Innovative Simulation for Health Care, IWISH* (pp. 78–89).

Dailey, K. W. (2003). *The lean manufacturing pocket handbook*. DW Publishing.

Dudek, P., Białoń, J., Piwowońska, J., Walczak, W., & Wrzała, K. (2023). The impact on the cost of making high pressure die castings with multi-cavity die and vacuum assistance. *Vacuum, 210*, 111859.

Herman, A., & Zikmund, P. (2010). The optimization of working cycles for HPDC technology. *Archives of Foundry Engineering, 10*(Issue Special 1/2010).

INCMA. (2004, April). Environmental management systems—Implementation guide for the foundry industry. *Indiana Cast Metal Association (INCMA)*. https://archive.epa.gov/sectors/web/pdf/foundry_complete.pdf

Longo, F., Nicoletti, L., Padovano, A., et al. (2019). Improving data consistency in industry 4.0: An application of digital lean to the maintenance record process. In *31st European Modeling and Simulation Symposium, EMSS 2019* (pp. 384–389).

PumpWorks Castings. (2004). *Top 10 casting safety considerations for foundries*. PumpWorks Casting. https://www.pumpworkscastings.com/casting-safety-considerations-for-foundries/

Rajkumar, I., Subash, K., Pradeesh, T. R., et al. (2021). Job safety hazard identification and risk analysis in the foundry division of a gear manufacturing industry. *Materials Today: Proceedings, 46*(Part 17), 7783–7788.

Saettaa, S., & Caldarellia, V. (2023). Lean and green: The green foundry simulation model. *Procedia Computer Science, 217*, 1622–1630.

Answers to MCQs and Selected Problems

Chapter 1

1.5-MCQs: (1) c, (2) a, (3) d, (4) a, (5) c, (6) b

Chapter 2

2.1-MCQs: (1) b, (2) a, (3) d, (4) c, (5) b, (6) b, (7) a
P-2.6. Fluidity = 41.388 in.
P-2.8. Heat energy = 81,648 J = 81.65 kJ
P-2.10. $Q = 651.9$ cm^3/s
P-2.12. $t = 7.29$ s

Chapter 3

3.1-MCQs: (1) d, (2) a, (3) b, (4) c, (5) a, (6) b, (7) c, (8) d, (9) a, (10) a
P-3.6.
Table P-3.6(c)

Cooling rate (K/s)	C_{min} (wt%)	C_{max} (wt%)	C_{av} (wt%)	C_{seg}
0.3	1.87	2.81	2.34	0.402
3.4	1.37	3.36	2.36	0.843
10.4	1.24	3.72	2.48	1.000

P-3.8. $C_m = 977,676$ s m^{-2}

P-3.10. $t_s = 4.7$ min

P-3.12. $t_s = 8.7$ min

Chapter 4

4.1-MCQs: (1) b, (2) d, (3) c, (4) a, (5) a, (6) b, (7) d

P-4.6. $R_L = 5.52$ mΩ

P-4.8. Monthly cost on electricity = \$ 8280.00

P-4.10. $m_{additive} = 132$ kg

P-4.12. $P_g = 404.412$ kW

P-4.14. $m_{scrap} = 21.134$ tons

Chapter 5

5.6-MCQs: (1) a, (2) d, (3) c, (4) b, (5) a, (6) c

P-5.8. (a) $D_{in} = 922$ mm, (b) $D_{out} = 948$ mm

P-5.10. $m_C = 13$ kg

P-5.12. Mass$_{coke} = 330$ kg

P-5.14. $D = 974$ mm

Chapter 6

6.1-MCQs: (1) b, (2) a, (3) d, (4) c, (5) b, (6) a, (7) d, (8) c, (9) a, (10) d

P-6.10. $P_n = 45.3$

P-6.12. 3%

Chapter 7

7.1-MCQs: (1) b, (2) d, (3) a, (4) c, (5) c, (6) a

P-7.8. $v = 1.253$ m/s

P-7.10. $T_{mf} = 23.3$ s

P-7.12. $R_e = 3747.3$, No air entrainment defect

P-7.14. $F_b = 11.77$ N

Chapter 9

9.1-MCQs: (1) b, (2) a, (3) c, (4) d, (5) a, (6) a, (7) d
P-9.10. $L_f = 5.01$ cm $= 5.1$ mm
P-9.12. $\sigma = 7.16$ MPa
P-9.14. $t_p = 11$ s

Chapter 10

10.1-MCQs: (1) c, (2) a, (3) b, (4) c, (5) a, (6) d, (7) b, (8) a
P-10.8. Runner p.a. + overflow p.a $= 200$ cm^2
P-10.10. MHR $= \$11.4$/h
P-10.12. USD 3.24/piece

Chapter 11

11.1-MCQs: (1) a, (2) d, (3) b, (4) c, (5) b, (6) a, (7) c, (8) b, (9) d, (10) b
P-11.8. Mass $= 1$ kg
P-11.10. $F = 647.46$ N
P-11.12. $N = 136$ rev/min
P-11.14. $n_2 = 1200$ rev/min
P-11.16. $L = 2.636$ m

Chapter 12

12.1-MCQs: (1) c, (2) a, (3) b, (4) d, (5) b, (6) a, (7) c, (8) d, (9) b, (10) c, (11) d, (12) b, (13) d, (14) b, (15) d, (16) c, (17) a, (18) d, (19) d, (20) a
P-12.8. $\sigma_{ut} = 1740$ MPa
P-12.10. 50HRC
P-12.12. KHN $\cong 996$
P-12.14. $S_e = 250$ MPa
P-12.16. $\theta_1 = 51°$

Chapter 13

13.1-MCQs: (1) b, (2) a, (3) d

Index

© The Editor(s) (if applicable) and The Author(s), under exclusive license to
Springer Nature Switzerland AG 2025
Z. Huda, *Metal Casting Engineering*, Mechanical Engineering Series,
https://doi.org/10.1007/978-3-031-84620-5